Bioinformatics

Bioinformatics

Arunima Mukhejee

Oxford Book Company

Jaipur ▪ New Delhi

Published By

Oxford Book Company

Regd. Off. "Bony Residency", Gate No. 2
Opp. Tilak Public School, Vishweswariya Nagar,
Gopalpura Road, Jaipur - 302018, India
Ph: 0141 - 2761280, Telefax: 2761381
email - oxfordbookjpr@gmail.com
website - oxfordbookcompany.com

Delhi Office: 102, 1st Floor, Satyam House,
4327/3, Ansari Road, Darya Ganj,
New Delhi - 110002, India
Ph: 011-45652440

Distributed By

KEYA INTERNATIONAL PRESS LLP
E-mail: keyainternationalpressllp@gmail.com

Published in India, 2024

ISBN : 9788189473518

Price: ₹ 1950.00

Typeset by Shivangi Computers, Jaipur

Printed and bound in India by Thomson Press Ltd.,
New Delhi

Preface

Bioinformatics is the analysis of biological information using computers and statistical techniques; the science of developing and utilising computer databases and algorithms to accelerate and enhance biological research. It encompasses the use of tools and techniques from three separate disciplines; molecular biology, computer science and the data analysis algorithms, which strictly define bioinformatics. The ultimate goal of bioinformatics is to uncover the wealth of biological information hidden in the mass of data and obtains a clearer insight into the fundamental biology of organisms. This new knowledge could have profound impacts on fields as varied as human health, agriculture, the environment, energy and biotechnology.

The present book provides an accessible introduction to a subject which has become a fundamental part of biological science today. It integrates an understanding of the biological background of bioinformatics with development of necessary computing skills. By discussing topics such as basics of genomics, principles of molecular genetics, computational molecular biology, protein sequencing, microarray data analysis, genome annotation and gene prediction techniques, it provides a comprehensive understanding of bioinformatics. The book will be an essential reading for students of genetics and bioinformatics.

Editor

Contents

1

Basics of Bioinformatics

Bioinformatics is the application of computational techniques to understand and organise the information associated with biological macromolecules. This uexpected union between the two subjects is largely attributed to the fact that life itself is an information technology; an organism's physiology is largely determined by its genes, which at its most basic can be viewed as digital information.

SCOPE OF BIOINFORMATICS

Bioinformatics has evolved into a full-fledged scientific discipline over the last decade. The definition of Bioinformatics is not restricted to computational molecular biology and computational structural biology. Bioinformatics uses advances in the area of computer science, information science, computer and information technology, communication technology to solve complex problems in life sciences like comparative genomics, structural genomics, transcriptiomics, Proteomics, cellunomics and metabolic pathway engineering. And particularly in biotechnology. Developments in these fields have direct implications to healthcare, medicine, discovery of next generation drugs, development of agricultural products, renewable energy, environmental protection etc.

Data capture, data warehousing and data mining have become major issues for biotechnologists and biological scientists due to sudden growth in quantitative data in biology such as complete genomes of biological species including human genome, protein sequences, protein 3-D structures, metabolic pathways databases, cell line & hybridoma information, biodiversity related information. Advancements in information technology, particularly the Internet, are being used to gather and access ever-increasing information in biology and biotechnology. Functional genomics, proteomics, discovery of new drugs and vaccines, molecular diagnostic kits and pharmacogenomics are some of the areas in which bioinformatics has become an integral part of Research & Development.

The knowledge of multimedia databases, tools to carry out data analysis and modelling of molecules and biological systems on computer workstations as well as in a network environment has become essential for any student of Bioinformatics. Bioinformatics, the multidisciplinary area, has grown so much that one divides it into molecular bioinformatics, organal bioinformatics and species bioinformatics. Issues related to biodiversity and environment, cloning of higher animals such as Dolly and Polly, tissue culture and cloning of plants have brought out that Bioinformatics is not only a support branch of science but is also a subject that directs future course of research in biotechnology and life sciences.

The importance and usefulness of Bioinformatics is realised in last few years by many industries. Therefore, large Bioinformatics R & D divisions are being established in many pharmaceutical companies, biotechnology companies and even in other conventional industry dealing with biological. Bioinformatics is thus rated as number one career in the field of biosciences. The need of trained manpower in this area is sharply on the rise but there are very few training institutions in the world where such training is provided. In short, Bioinformatics deals with database creation, data analysis and modelling.

Data capturing is done not only from printed material but also from network resources. Databases in biology are generally in the multimedia form organised in relational database model. Modelling is done not only on single biological molecule but also on multiple systems thus requiring a use of high performance computing systems.

POTENTIAL OF BIOINFORMATICS

The potential of bioinformatics in the identification of useful genes leading to the development of new gene products, drug discovery and drug development has led to a paradigm shift in biology and biotechnology-these fields are becoming more & more computationally intensive. The new paradigm, now emerging, is that all the genes will be known "in the sense of being resident in database available electronically", and the starting point of biological investigation will be theoretical and a scientist will begin with a theoretical conjecture and only then turning to experiment to follow or test the hypothesis. With a much deeper understanding of the biological processes at the molecular level, the Bioinformatics scientist have developed new techniques to analyse genes on an industrial scale resulting in a new area of science known as 'Genomics'.

The shift from gene biology has resulted in the development of strategies-from lab techniques to computer programs to analyse whole batch of genes at once. Genomics is revolutionising drug development, gene therapy, and our entire approach to health care and human medicine. The genomic discoveries are getting translated in to practical biomedical results through Bioinformatics applications. Work on proteomics and genomics will continue using highly sophisticated software tools and data networks that can carry multimedia databases. Thus, the research will be in the development of multimedia databases in various areas of life sciences and biotechnology. There will be an urgent need for development of software tools for data mining, analysis and modelling, and downstream processing.

Security of data, data transfer and data compression, auto checks on data accuracy and correctness will also be major research area of bioinformatics. The use of virtual Reality in drug design, metabolic pathway design, and unicellular organism design, paving the way to design and modification of unicellular organisms, will be the challenges, which are the challenges, which Bioinformatics scientist and specialist have to tackle. It has now been universally recognised that Bioinformatics is the key to the new grand data-intensive molecular biology that will take us into 21 century

AIMS OF BIOINFORMATICS

The aims of bioinformatics are threefold:

First, at its simplest bioinformatics organises data in a way that allows researchers to access existing information and to submit new entries as they are produced, eg the Protein Data Bank for 3D macromolecular structures. While data-curation is an essential task, the information stored in these databases is essentially useless until analysed. Thus the purpose of bioinformatics extends much further.

The second aim is to develop tools and resources that aid in the analysis of data. For example, having sequenced a particular protein, it is of interest to compare it with previously characterised sequences. This needs more than just a simple text-based search and programmes such as FASTA and PSI-BLAST must consider what comprises a biologically significant match. Development of such resources dictates expertise in computational theory as well as a thorough understanding of biology.

The third aim is to use these tools to analyse the data and interpret the results in a biologically meaningful manner. Traditionally, biological studies examined individual systems in detail, and frequently compared them with a few that are related. In bioinformatics, we can now conduct global analyses of all the available data with the aim of uncovering common principles that apply across many systems and highlight novel features.

Sources of Information

The sources of information may be divided into raw DNA sequences, protein sequences, macromolecular structures, genome sequences, and other whole genome data. Raw DNA sequences are strings of the four baseletters comprising genes, each typically 1,000 bases long. The GenBank repository of nucleic acid sequences currently holds a total of 9.5 billion bases in 8.2 million entries. At the next level are protein sequences comprising strings of 20 amino acid-letters. At present there are about 300,000 known protein sequences, with a typical bacterial protein containing approximately 300 amino acids. Macromolecular structural data represents a more complex form of information. There are currently 13,000 entries in the Protein Data Bank, PDB, most of which are protein structures. A typical PDB file for a medium-sized protein contains the xyz coordinates of approximately 2,000 atoms.

Scientific euphoria has recently centred on whole genome sequencing. As with the raw DNA sequences, genomes consist of strings of baseletters, ranging from 1.6 million bases in Haemophilus influenzae to 3 billion in humans. An important aspect of complete genomes is the distinction between coding regions and noncoding regions—'junk' repetitive sequences making up the bulk of base sequences especially in eukaryotes. We can now measure expression levels of almost every gene in a given cell on a whole-genome level although public availability of such data is still limited.

Expression level measurements are made under different environmental conditions, different stages of the cell cycle and different cell types in multicellular organisms. Currently the largest dataset for yeast has made approximately 20 time-point measurements for 6,000 genes. Other genomicscale data include biochemical information on metabolic pathways, regulatory networks, protein-protein interaction data from two-hybrid experiments, and systematic knockouts of individual genes to test the viability of an organism.

What is apparent from this list is the diversity in the size and complexity of different datasets. There are invariably more sequence-based data than structural data because of the relative ease with which they can be produced. This is partly related to the greater complexity and information-content of individual structures compared to individual sequences. While more biological information can be derived from a single structure than a protein sequence, the lack of depth in the latter is remedied by analysing larger quantities of data.

Redundancy and Multiplicity

A concept that underpins most research methods in bioinformatics is that much of this data can be grouped together based on biologically meaningful similarities. For example, sequence segments are often repeated at different positions of genomic DNA. Genes can be clustered into those with particular functions (eg enzymatic actions) or according to the metabolic pathway to which they belong, although here, single genes may actually possess several functions.

Going further, distinct proteins frequently have comparable sequences—organisms often have multiple copies of a particular gene through duplication while different species have equivalent or similar proteins that were inherited when they diverged from each other in evolution. At a structural level, we predict there to be a finite number of different tertiary structures—estimates range between 1,000 and 10,000 folds and proteins adopt equivalent structures even when they differ greatly in sequence. As a result, although the number of structures in the PDB has increased exponentially, the rate of discovery of novel folds has actually decreased.

There are common terms to describe the relationship between pairs of proteins or the genes from which they are derived: analogous proteins have related folds, but unrelated sequences, while homologous proteins are both sequentially and structurally similar. The two categories can sometimes be difficult to distinguish especially if the relationship between the two proteins is remote. Among homologues, it is useful to distinguish between orthologues, proteins in different species that have evolved from a common ancestral gene, and paralogues, proteins that are related by gene duplication within a genome. Normally, orthologues retain the same function while paralogues evolve distinct, but related functions.

An important concept that arises from these observations is that of a finite "parts list" for different organisms : an inventory of proteins contained within an organism, arranged according to different properties such as gene sequence, protein fold or function. Taking protein folds as an example, the tertiary structures of proteins adopt one of a limited repertoire of folds. As the number of different fold families is considerably smaller than the number of gene families, categorising the proteins by fold provides a substantial simplification of the contents of a genome.

Similar simplifications can be provided by other attributes such as protein function. As such, we expect this notion of a finite parts list to become increasingly common in the future genomic analyses. The following major databases that provide access to the primary sources of information, and also introduce some secondary databases that systematically group the data. These classifications ease comparisons between genomes and their products, allowing the identification of common themes between those that are related and highlighting features that are unique to some.

Protein Sequence Databases

Protein sequence databases are categorised as primary, composite or secondary. Primary databases contain over 300,000 protein sequences and function as a repository for the raw data. Some more common repositories, such as SWISS-PROT and PIRInternational, annotate the sequences as well as describe the proteins' functions, its domain structure and post-translational modifications. Composite databases such as OWL and the NRDB compile and filter sequence data from different primary databases to produce combined non-redundant sets that are more complete than the individual databases and also include protein sequence data from the translated coding regions in DNA sequence databases.

Secondary databases contain information derived from protein sequences and help the user determine whether a new sequence belongs to a known protein family. One of the most popular is PROSITE, a database of short sequence patterns and profiles that characterise biologically significant sites in proteins. PRINTS expands on this concept and provides a compendium of protein fingerprints -groups of conserved motifs that characterise a protein family. Motifs are usually separated along a protein sequence, but may be contiguous in 3D-space when the protein is folded.

By using multiple motifs, fingerprints can encode protein folds and functionalities more flexibly than PROSITE. Finally, Pfam contains a large collection of multiple sequence alignments and profile Hidden Markov Models covering many common protein domains. Pfam-A comprises accurate manually compiled alignments while Pfam-B is an automated clustering of the whole SWISS-PROT database. These different secondary databases have recently been incorporated into a single resource named InterPro.

Structural Databases

The Protein Data Bank, PDB, provides a primary archive of all 3D structures for macromolecules such as proteins, RNA, DNA and various complexes. Most of the ~13,000 structures are solved by x-ray crystallography and NMR, but some theoretical models are also included. As the information provided in individual PDB entries can be difficult to extract, PDBsum provides a separate Web page for every structure in the PDB displaying detailed structural analyses, schematic diagrams and data on interactions between different molecules in a given entry. Three major databases classify proteins by structure in order to identify structural and evolutionary relationships: CATH, SCOP, and FSSP databases. All comprise hierarchical structural taxonomy where groups of proteins increase in similarity at lower levels of the classification tree. In addition, numerous databases focus on particular types of macromolecules. These include the Nucleic Acids Database, NDB, for structures related to nucleic acids, the HIV protease database for HIV-1, HIV-2 and SIV protease structures and their complexes, and ReLiBase for receptor-ligand complexes.

Nucleotide and Genome Sequences

The biggest excitement currently lies with the availability of complete genome sequences for different organisms. The GenBank, EMBL and DDBJ databases contain DNA sequences for individual genes that encode protein and

RNA products. Much like the composite protein sequence database, the Entrez nucleotide database compiles sequence data from these primary databases. As whole-genome sequencing is often conducted through international collaborations, individual genomes are published at different sites. The Entrez genome database brings together all complete and partial genomes in a single location and currently represents over 1,000 organisms.

In addition to providing the raw nucleotide sequence, information is presented at several levels of detail including: a list of completed genomes, all chromosomes in an organism, detailed views of single chromosomes marking coding and non-coding regions, and single genes. At each level there are graphical presentations, precomputed analyses and links to other sections of Entrez. For example, annotations for single genes include the translated protein sequence, sequence alignments with similar genes in other genomes and summaries of the experimentally characterised or predicted function.

GeneCensus also provides an entry point for genome analysis with an interactive wholegenome comparison from an evolutionary perspective. The database allows building of phylogenetic trees based on different criteria such as ribosomal RNA or protein fold occurrence. The site also enables multiple genome comparisons, analysis of single genomes and retrieval of information for individual genes. The most straightforward application of the database is to predict the function of uncharacterised proteins through their homology to characterised proteins, and also to identify phylogenetic patterns of protein occurrence.

Gene Expression Data

A most recent source of genomicscale data has been from expression experiments, which quantify the expression levels of individual genes. These experiments measure the amount of mRNA or protein products that are produced by the cell. For the former, there are three main technologies:

— the cDNA microarray,

— Affymatrix GeneChip and

— SAGE methods.

The first method measures relative levels of mRNA abundance between different samples, while the last two measure absolute levels. Most of the effort in gene expression analysis has concentrated on the yeast and human genomes and as yet, there is no central repository for this data. For yeast, the Young, Church and Samson datasets use the GeneChip method, while the Stanford cell cycle, diauxic shift and deletion mutant datasets use the microarray. Most measure mRNA levels throughout the whole yeast cell cycle, although some focus on a particular stage in the cycle.

For humans, the main application has been to understand expression in tumour and cancer cells. The Molecular Portraits of Breast Tumours, Lymphoma and Leukaemia Molecular Profiling projects provide data from microarray experiments on human cancer cells. The technologies for measuring protein abundance are currently limited to. 2D gel electrophoresis followed by mass spectrometry. As gels can only routinely resolve about 1,000 proteins, only the most abundant can be visualised. At present, data from these experiments are only available from the literature.

Integration of Multiple Source of Data

The most profitable research in bioinformatics often results from integrating multiple sources of data. For instance, the 3D coordinates of a protein are more useful if combined with data about the protein's function, occurrence in different genomes, and interactions with other molecules. In this way, individual pieces of information are put in context with respect to other data. Unfortunately, it is not always straightforward to access and cross-reference these sources of information because of differences in nomenclature and file formats.

At a basic level, this problem is frequently addressed by providing external links to other databases, for example in PDBsum, web-pages for individual structures direct the user towards corresponding entries in the PDB, NDB, CATH, SCOP and SWISS-PROT. At a more advanced level, there have been efforts to integrate access across several data sources. One is the Sequence Retrieval System, SRS, which allows flatfile databases to be indexed to each other; this allows the user to retrieve, link and access entries from nucleic acid, protein sequence, protein motif, protein structure and bibliographic databases.

UNDERSTANDING BIOLOGICAL DATA

Another is the Entrez facility, which provides similar gateways to DNA and protein sequences, genome mapping data, 3D macromolecular structures and the PubMed bibliographic database. A search for a particular gene in either database will allow smooth transitions to the genome it comes from, the protein sequence it encodes, its structure, bibliographic reference and equivalent entries for all related genes.

Investigations of structural data include prediction of secondary and tertiary protein structures, producing methods for 3D structural alignments,examining protein geometries using distance and angular measurements, calculations of surface and volume shapes and analysis of protein interactions with other subunits, DNA, RNA and smaller molecules. These studies have lead to molecular simulation topics in which structural data are used to calculate the energetics involved in stabilising macromolecular structures, simulating movements within macromolecules, and computing the energies involved in molecular docking.

The increasing availability of annotated genomic sequences has resulted in the introduction of computational genomics and proteomics—large-scale analyses of complete genomes and the proteins that they encode. Research includes characterisation of protein content and metabolic pathways between different genomes, identification of interacting proteins, assignment and prediction of gene products, and large-scale analyses of gene expression levels. Some of these research topics will be demonstrated in our example analysis of transcription regulatory systems.

In addition to finding relationships between different proteins, much of bioinformatics involves the analysis of one type of data to infer and understand the observations for another type of data. An example is the use of sequence and structural data to predict the secondary and tertiary structures of new protein sequences. These methods, especially the former, are often based on statistical rules derived from structures, such as the propensity for certain amino acid sequences to produce different secondary structural elements. Another example is the use of structural data to understand a protein's function; here studies have investigated the relationship different protein folds and their functions and analysed similarities between different binding sites in the absence of homology. Combined with similarity measurements, these studies provide us with an understanding of how much biological information can be accurately transferred between homologous proteins.

INFORMATION TECHNIQUES

The distinct subject areas require different types of informatics techniques. Briefly, for data organisation, the first biological databases were simple flat files. However with the increasing amount of information, relational database methods with Web-page interfaces have become increasingly popular. In sequence analysis, techniques include string comparison methods such as text search and 1-dimensional alignment algorithms. Motif and pattern identification for multiple sequences depend on machine learning, clustering and data-mining techniques.

3D structural analysis techniques include Euclidean geometry calculations combined with basic application of physical chemistry, graphical representations of surfaces and volumes, and structural comparison and 3D matching methods. For molecular simulations, Newtonian mechanics, quantum mechanics, molecular mechanics and electrostatic calculations are applied. In many of these areas, the computational methods must be combined with good statistical analyses in order to provide an objective measure for the significance of the results.

Transcription Regulation

DNA-binding proteins have a central role in all aspects of genetic activity within an organism, participating in processes such as transcription, packaging, rearrangement, replication and repair. Through this example, we demonstrate how bioinformatics has been used to increase our knowledge of biological systems and also illustrate the practical applications of the different subject areas.

Later, several genomic studies that have characterised the nature of transcription factors in different organisms, and the methods that have been used to identify regulatory binding sites in the upstream regions. Finally, an overview of gene expression analyses that have been recently conducted and suggest future uses of transcription regulatory analyses to rationalise the observations made in gene expression experiments.

APPLICATIONS OF BIOINFORMATICS

In the following, we can see some of the major uses of bioinformatics.

Finding Homologues

One of the driving forces behind bioinformatics is the search for similarities between different biomolecules. Apart from enabling systematic organisation of data, identification of protein homologues has some direct practical uses. The most obvious is transferring information between related proteins. For example, given a poorly characterised protein, it is possible to search for homologues that are better understood and with caution, apply some of the knowledge of the latter to the former.

Specifically with structural data, theoretical models of proteins are usually based on experimentally solved structures of close homologues. Similar techniques are used in fold recognition in which tertiary structure predictions depend on finding structures of remote homologues and checking whether the prediction is energetically viable. Where biochemical or structural data are lacking, studies could be made in low-level organisms like yeast and the results applied to homologues in higher-level organisms such as humans, where experiments are more demanding.

An equivalent approach is also employed in genomics. Homologuefinding is extensively used to confirm coding regions in newly sequenced genomes and functional data is frequently transferred to annotate individual genes. On a larger scale, it also simplifies the problem of understanding complex genomes by analysing simple organisms first and then applying the same principles to more complicated ones—this is one reason why early structural genomics projects focused on Mycoplasma genitalium.

Ironically, the same idea can be applied in reverse. Potential drug targets are quickly discovered by checking whether homologues of essential microbial proteins are missing in humans. On a smaller scale, structural differences between similar proteins may be harnessed to design drug molecules that specifically bind to one structure but not another.

Rational Drug Design

One of the earliest medical applications of bioinformatics has been in aiding rational drug design. FGiven the nucleotide sequence, the probable amino acid sequence of the encoded protein can be determined using translation software. Sequence search techniques can then be used to find homologues in model organisms, and based on sequence similarity, it is possible to model the structure of the human protein on experimentally characterised structures. Finally, docking algorithms could design molecules that could bind the model structure, leading the way for biochemical assays to test their biological activity on the actual protein.

Large-scale Censuses

Although databases can efficiently store all the information related to genomes, structures and expression datasets, it is useful to condense all this information into understandable trends and facts that users can readily understand. Broad generalisations help identify interesting subject areas for further detailed analysis, and place new observations in a proper context. This enables one to see whether they are unusual in any way.

Through these large-scale censuses, one can address a number of evolutionary, biochemical and biophysical questions. For example, are specific protein folds associated with certain phylogenetic groups? How common are different folds within particular organisms? And to what degree are folds shared between related organisms? Does this extent of sharing parallel measures of relatedness derived from traditional evolutionary trees? Initial studies show that the frequency of folds differs greatly between organisms and that the sharing of folds between organisms does in fact follow traditional phylogenetic classifications.

We can also integrate data on protein functions; given that the particular protein folds are often related to specific biochemical functions, these findings highlight the diversity of metabolic pathways in different organisms. One of the most exciting new sources of genomic information is the expression dat a.

Combining expression information with structural and functional classifications of proteins we can ask whether the high occurrence of a protein fold in a genome is indicative of high expression levels. Further genomic scale data that we can consider in large-scale surveys include the subcellular localisations of proteins and their interactions with each other. In conjunction with structural data, we can then begin to compile a map of all proteinprotein interactions in an organism.

Applications in Medical Sciences

Most recent applications in the medical sciences have centred on gene expression analysis. This usually involves compiling expression data for cells affected by different diseases, eg cancer and ateriosclerosis, and comparing the measurements against normal expression levels. Identification of genes that are expressed differently in affected cells provides a basis for explaining the causes of illnesses and highlights potential drug targets. Given a lead compound, microarray experiments can then be used to evaluate responses to pharmacological intervention, and also provide early tests to detect or predict the toxicity of trial drugs.

Further advances in bioinformatics combined with experimental genomics for individuals are predicted to revolutionalise the future of healthcare. A typical scenario for a patient may start with post-natal genotyping to assess susceptibility or immunity from specific diseases and pathogens. With this information, a unique combination of vaccines could be prescribed, minimising the healthcare costs of unnecessary treatments and anticipating the onslaught of diseases later in life.

Regular lifetime screenings could lead to guidance for nutrition intake and early detections of any illnesses. In addition, drug-based treatments could be tailored specifically to the patient and disease, thus providing the most effective course of medication with minimal side-effects. Given the present rate of development, such a scenario in healthcare appears to be possible in the not too distant future.

HISTORY OF GENOMICS

The recent history of genomics also has been driven by technological advances. Foremost among these advances were the methodologies of the polymerase chain reaction (PCR) and automated DNA sequencing. PCR methods allowed the ampli.cation of usable amounts of DNA from very small amounts of starting material. Automated DNA sequencing methods have progressed to the point that today the entire DNA sequence of microbial genomes containing several million base pairs can be obtained in less than one week. These accomplishments set the stage for the human genome project.

As early as 1984 the small genomes of several.microbes and bacteriophages had been mapped and partially sequenced; however, the modern era of genomics was not formally initiated until 1986 at an international conference in Santa Fe, New Mexico sponsored by the Office of Health and Environmental Research of the US Department of

Energy. At this meeting, the desirability and feasibility of implementing a human genome program was unanimously endorsed by leading scientists from around the world. This meeting led to a 1988 study by the National Research Council titled Mapping and Sequencing the Human Genome that recommended the United States support a human genome program and presented an outline for a multiphase plan.

In that same year, three genome research centers were established at the Lawrence Berkeley, Lawrence Livermore, and Los Alamos national laboratories. At the same time, under the leadership of Director James Wyngaarden, the National Institutes of Health established the Office of Genome Research which in 1989 became the National Center for Human Genome Research, directed by James D. Watson. The next ten years witnessed rapid progress and technology developments in automated sequencing methods. These technologies led to the establishment of largescale DNA sequencing projects at many public research institutions around the world such as the Whitehead Institute in Boston, MA and the Sanger Centre in Cambridge, UK. These activities were accompanied by the rapid development of computational and informational methods to meet challenges created by an increasing .ow of data from large-scale genome sequencing projects.

In 1991 Craig Venter at the National Institutes of Health developed a way of finding human genes that did not require sequencing of the entire human genome. He relied on the estimate that only about 3 percent of the genome is composed of genes that express messenger RNA. Venter suggested that the most efficient way to find genes would be to use the processing machinery of the cell. At any given time, only part of a cell's DNA is transcriptionally active. These "expressed" segments of DNA are converted and edited by enzymes into mRNA molecules. Using an enzyme, reverse transcriptase, cellular mRNA fragments can be transcribed into complementary DNA (cDNA). These stable cDNA fragments are called expressed sequence tags, or ESTs.

Computer programs that match overlapping ends of ESTs were used to assemble these cDNA sequences into longer sequences representing large parts, or all, of many human genes. In 1992, Venter left NIH to establish The Institute for Genomic Research, TIGR. By 1995 researchers in public and private institutions had isolated over 170000 ESTs, which were used to identify more than half of the then estimated 60000 to 80000 genes in the human genome. In 1998, Venter joined with Perkin-Elmer Instruments to form Celera Genomics. With the end in sight, in 1998 the Human Genome Program announced a plan to complete the human genome sequence by 2003, the 50th anniversary of Watson and Crick's description of the structure of DNA. The goals of this plan were to: Achieve coverage of at least 90% of the genome in a working draft based on mapped clones by the end of 2001.

— Finish one-third of the human DNA sequence by the end of 2001.

— Finish the complete human genome sequence by the end of 2003.

— Make the sequence totally and freely accessible.

On June 26, 2000, President Clinton met with Francis Collins, the Director of the Human Genome Program, and Craig Venter of Celera Genomics to announce that they had both completed "working drafts" of the human genome, nearly two years ahead of schedule. These drafts were published in special issues of the journals Science and Nature early in 2001 and the sequence is online at the National Center for Biotechnology Information (NCBI) of the Library of Medicine at the National Institutes of Health. The NCBI databases also contain complete or in progress genomic sequences for ten Archaea and 151 bacteria as well as the genomic sequences of eight eukaryotes including: the parasites Leishmania major and Plasmodium falciparum; the worm Caenorhabditis elegans; the yeast Saccharomyces cerevisiae; the fruit fly Drosophila melanogaster; the mouse Mus musculus; and the plant Arabidopsis thaliana.

Many more genome sequencing projects are under way in private and public research laboratories that are not yet available on public databases. It is anticipated that the acquisition of new genome sequence data will continue to accelerate. This exponential increase in DNA sequence data has fuelled a drive to develop technologies and computational methods to use this information to study biological problems at levels of complexity never before possible.

DNA Array Formats

Array technologies monitor the combinatorial interaction of a set of molecules, such as DNA fragments and proteins, with a predetermined library of molecular probes. The currently most advanced of these technologies is the use of DNA arrays, also called DNA chips, for simultaneously measuring the level of the mRNA gene products of a living cell. In its most simple sense, a DNA array is defined as an orderly arrangement of tens to hundreds of thousands of unique DNA molecules (probes) of known sequence. There are two basic sources for the DNA probes on an array. Either each unique probe is individually synthesised on a rigid surface (usually glass), or pre-synthesised probes (oligonucleotides or PCR products) are attached to the array platform (usually glass or nylon membranes). The various types of DNA arrays currently available for gene expression profiling, as well as some developing technologies, are summarised here.

In Situ Synthesised Oligonucleotide Arrays

The first in situ probe synthesis method for manufacturing

DNA arrays was the photolithographic method developed by Fodor et al. and commercialised by Affymetrix Inc. First, a set of oligonucleotide DNA probes (each 25 or so nucleotides in length) is defined based on its ability to hybridise to complementary sequences in target genomic loci or genes of interest. With this information, computer algorithms are used to design photolithographic masks for use in manufacturing the probe arrays. Selected addresses on a photo-protected glass surface are illuminated through holes in the photolithographic mask, the glass surface is flooded with the first nucleotide of the probes to be synthesised at the selected addresses, and photo-chemical coupling occurs at these sites.

For example, the addresses on the glass surface for all probes beginning with guanosine are photo-activated and chemically coupled to guanine bases. This step is repeated three more times with masks for all addresses with probes beginning with adenosine, thymine, or cytosine. The cycle is repeated with masks designed for adding the appropriate second nucleotide of each probe. During the second cycle, modified phosphoramidite moieties on each of the nucleosides attached to the glass surface in the first step are light-activated through appropriate masks for the addition of the second base to each growing oligonucleotide probe. This process is continued until unique probe oligonucleotides of a defined length and sequence have been synthesised at each of thousands of addresses on the glass surface (Figure 1).

Several companies such as Protogene (Menlo Park, CA) and Agilent Technologies (Palo Alto, CA) in collaboration with Rosetta Inpharmatics (Kirkland,WA) of Merck & Co. Inc. (Whitehouse Station, NJ) have developed in situ DNA array platforms through proprietary modifications of a standard piezoelectric (ink-jet) printing process that unlike the manufacturing process for Affymetrix GeneChips™, does not require photolithography. These in situ synthesised oligonucleotide arrays are fabricated directly on a glass support on which oligonucleotides up to 60 nucleotides are synthesised using standard phosphoramidite chemistry. The ink-jet printing technology is capable of depositing very small volumes—picoliters per spot—of DNA solutions very rapidly and very accurately. It also delivers spot shape uniformity that is superior to other deposition methods.

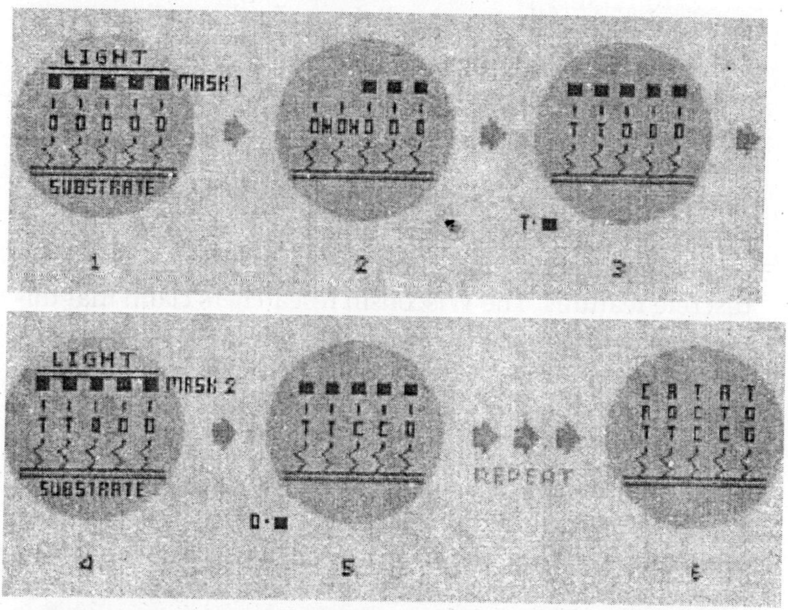

Figure 1. The Affymetrix method for the manufacture of in situ synthesised DNA microarrays (1) A photo-protected glass substrate is selectively illuminated by light passing through a photolithographic mask. (2) Deprotected areas are activated. (3) The surface is flooded with a nucleoside solution and chemical coupling occurs at photo-activated positions. (4) A new photolithographic mask pattern is applied. (5) The coupling step is repeated. (6) This process is repeated until the desired set of probes is obtained.

Researchers in the Nano-fabrication Center at the University of Wisconsin have developed yet another method for the manufacture of in situ synthesised DNA arrays that also does not require photolithographic masks. This technology known as MAS for maskless array synthesiser capitalises on existing electronic chips used in overhead projection known as digital light processors (DLPs).

A DLP is an array of up to 500000 tiny aluminum mirrors arranged on a computer chip. By electronic manipulation of the mirrors, light can be directed to specific addresses on the surface of a DNA array substrate, thus eliminating the need for expensive photolithographic masks. This technology is being implemented by NimbleGen Systems, LLC (Madison, WI). DNA arrays containing over 307000 discrete features are currently being synthesised and plans are under way to synthesise a second-generation MAS array containing over 2 million discrete features. The Wisconsin researchers claim that this method will greatly reduce the time and cost for the manufacture of high-density in situ synthesised DNA mircoarrays, and bring this activity into individual research laboratories.

CombiMatrix (Snoqualmie, WA) and Nanogen (San Diego, CA) are developing electrical addressing systems for the manufacture of DNA arrays on semiconductor chips. The CombiMatrix method involves attaching each addressable site on the chip to an electrical conduit (electrode) applied over a layer of porous material. Each DNA probe is synthesised one base at a time by flooding the porous layer with a nucleoside and activating each electrode where a new base is to be added. Once activated, the electrode causes an electrochemical reaction to occur which produces chemicals that react with the existing nucleotides, or chains of DNA, at that site for bonding to the probe site or to the next nucleotide base.

At present, CombiMatrix has produced DNA arrays with 100 μm features that possess 1024 test sites within less than a square centimeter. Researchers at CombiMatrix believe that by using a standard 0.25-µm semiconductor fabrication process, they can produce a biological array processor with over 1000000 sites per square centimeter. Nanogen uses a similar process to attach pre-synthesised oligonucleotides to electronically addressable sites on a semiconductor chip. To date, Nanogen has only produced a 99 probe array suitable for forensic and diagnostic purposes; however, Nanogen's researchers anticipate electronic arrays with thousands of addresses for genomics applications.

Pre-synthesised DNA Arrays

The method of attaching pre-synthesised DNA probes (usually 100-5000 bases long) to a solid surface such as glass (or nylon filter) supports was conceived 25 years ago by Ed Southern and more recently popularised by the Patrick O. Brown laboratory at Stanford University. While the early manufacturing methods for miniaturised DNA arrays using in situ probe synthesis required sophisticated and expensive robotic equipment, the glass slide DNA array manufacturing methods of Brown made DNA arrays affordable for academic research laboratories. As early as 1996 the Brown laboratory published step-by-step plans for the construction of a robotic DNA arrayer on the internet. Since that time, many commercial DNA arrayers have become available. Besides the commercially produced Affymetric GeneChips', these Brown-type glass slide DNA arrays are currently the most popular format for gene expression profiling experiments.

The Brown method for printing glass slide DNA arrays involves the robotic spotting of small volumes (in the nanoliter to picoliter range) of a DNA probe sample onto a 25 × 76 × 1 mm glass slide surface previously coated with poly-lysine or poly-amine for electrostatic adsorption of the DNA probes onto the slide. Depending upon the pin type and the exact printing technology employed, 200 to 10000 spots ranging in size from 500 to 75 μm can be spotted in a 1-cm^2 area. Many public and private research institutions in the USA and abroad have developed core facilities for the in-house manufacture of custom glass slide DNA arrays.

ISSUES AND FUTURE TRENDS IN BIOINFORMATICS

The greatest challenge facing the molecular biology community today is to make sense of the wealth of data that has been produced by the genome sequencing projects. Traditionally, molecular biology research was carried out entirely at the experimental laboratory bench but the huge increase in the scale of data being produced in this genomic era has seen a need to incorporate computers into this research process. Sequence generation, and its subsequent storage, interpretation and analysis are entirely computer dependent tasks. However, the molecular biology of an organism is a very complex issue with research being carried out at different levels including the genome, proteome, transcriptome and metabalome levels.

The first challenge facing the bioinformatics community today is the intelligent and efficient storage of this mass of data. It is then their responsibility to provide easy and reliable access to this data. The data itself is meaningless before analysis and the sheer volume present makes it impossible for even a trained biologist to begin to interpret it manually. Therefore, incisive computer tools must be developed to allow the extraction of meaningful biological information. There are three central biological processes around which bioinformatics tools must be developed:

— DNA sequence determines protein sequence
— Protein sequence determines protein structure
— Protein structure determines protein function

The integration of information learned about this key biological process should allow us to achieve the long-term goal of the complete understanding of the biology of organisms.

Structure Prediction of Protein

The masses of data led to numerous challenging bioinformatics problems. The protein structure prediction problems aims to predict the 3D structure of a protein from its protein seqeunce. It is one of the most important unsolved problems of computational biology, which was shown to be at least NP-complete. For this reason, simplified models have been successfully used by several groups in hierarchical approaches for protein folding. The most important class of simplified models are the lattice models, where monomers are represented using a unified size having positions in a regular lattice. They are used for structure prediction as well as a method for investigating general properties of protein folding.

There is a bunch of groups working with lattice proteins. Examples of how lattice proteins can be used for predicting the native structure or for investigating principles of protein folding are. Most of them use heuristic methods, ranging from Monte-Carlo simulated annealing to genetic algorithms, purely heuristic methods like hydrophobic zipper and the chain growth algorithm, as well as complete enumeration.

First steps have been made to improve the situation on the algorithmic part. The first improvement was the introduction of an exact algorithm for finding minimal energy conformations in the cubic HP-Model. The algorithm is called CHCC for (Constraint Hydrophobic Core Construction). This approach works only for the cubic lattice, but not for the FCC-lattice which is much better suited for modelling real protein conformations.

The second improvement is the appearance of a bunch of approximation algorithms for different lattice models. Although they are a very important first step in the right direction, their approximation ratio is still not good enough to be used in practice.

Another interesting approach is to combine secondary structure predictions with lattice models. The secondary structure consists of local structure motifs like α-helices and β-sheets. One searches for conformations in the lattice model with low energy having the predicted secondary structure elements. Here, major improvements have been achieved using again a constraint-based approach and the FCC lattice.

Alignment Methods

The alignment problem is a very important problem in bioinformatics, where one searches for an alignment of two strings, optimising a certain score. The strings represent biological macromolecules as DNAs or proteins. The scoring scheme evaluates the aligned columns. Alignment is one of the core techniques, which are used to search for similar objects in biology. Initially, they were used only for sequences, but it was later extended to many other problems (protein structure, pathways etc.)

In the simplest case, alignment is identical to computing the edit distance of strings. This problem is usually solved by dynamic programming (DP) as e.g. by Needleman and Wunsch in. However, dynamic programming approaches suffer from their inflexibility. If the problem is slightly modified, one has to develop a new DP algorithm (if one exists at all). For example, it is an unsolved problem to align two sequences incorporating biological knowledge that tells us which sub-sequences/domains should be aligned.

In addition, many extensions of the basic alignment problem have been investigated. Examples of this kind of "extended basic alignment methods" is RNA-Sequence-Structure Alignment. This kind of sequence/structure similarity has been extended to proteins in many different ways, where the problem becomes NP-complete. This are protein threading, the contact map problem and the alignment of TOPS-diagrams.

Metabolic and Signalling Networks

In recent years, molecular biology has engaged in a large-scale effort to elucidate cellular processes in terms of their biochemical basis at the molecular level. Mass production of post genomic experimental results, such as mRNA expression data, protein expression or protein-protein interaction data, is following and completing the initial piecemeal catalogue of elementary components—genes and proteins—of the sequencing and genomic analyses projects by progressively painting a global picture of the complex interactions that take place in a cell. Exploiting these experimental data to understand the underlying processes requires much more than database integration and storage: it calls for a strong parallel effort on the formal representation of biological processes.

Several formalisms have been proposed in recent years for the modelling of metabolic pathways, extracellular and intracellular signalling pathways, or gene regulatory networks: boolean networks, ordinary differential equations, and more recently hybrid Petri nets and hybrid automata. Formal concurrent languages were also considered, including hybrid concurrent constraint languages, or rewriting logics. Regev and Shapiro were the first to propose the use of the π-calculus.

The current state-of-the-art in modelling is mostly based on simulation and graphical display, with some attempts towards stability and bifurcation analyses of dynamical behaviour on small systems described either by differential equations or by discretisations thereof. A different approach promotes symbolic manipulation and exploration of the model by means of computational logics which are commonplace, in hardware verification for instance. Formal methods extend the ways one can play with a given model and thus may second simulation and even replace it when quantitative information is sparse or inaccurate.

BIOINFORMATICS AND THE WEB

There are masses of bioinformatics data and algorithms working on it. Much of the data and tools are published online. Bioinformatics could play the role for the Semantic Web, which physics played for the web. The reasons are manyfold:

— There are masses of information
— The data are public
— The data are online
— The data are (more and more often) published in XML
— Data standards are accepted and actively developed by the bioinformatics community
— Much valuable information is scattered
— Systems integration and interoperation are a prime concern
— One can predict that in the not too distant future many tools and databases will be accessible as web services

BIOINFORMATICS DATA

With much data online accessible, let us consider some of the data sources. In a number of the most used and best known databanks were studied: GenBank, EMBL, DDBJ, SWISS-PROT, PIR, ENZYME, PDB, MMDB, PROSITE, PRINTS and BLOCKS. For these systems we investigated the organisation of the data, the data content and the data retrieval possibilities. With respect to the organisation of the data we looked at the kind of data, the source, the data model, the update frequency and the location of the data. The chosen databanks store information about nucleotide sequences (GenBank, EMBL, DDBJ), protein sequences (SWISS-PROT, PIR, ENZYME), 3D macromolecular structures (PDB, MMDB) and protein families (PROSITE, PRINTS, BLOCKS).

DATABANKS INTEGRATION

There exist a number of systems that provide access to multiple biological databanks. These systems can be divided into two categories. The category of the linkdriven federations contains most of the currently used websites that provide an interface to multiple biological resources, such as Entrez and SRS. These systems support a number of basic queries via a web interface. Often, they also allow to use alignment algorithms such as BLAST on the result of a query. Usually, the users need to explicitly state which resources should be used for retrieving the answers, requiring good knowledge of the underlying sources.

The data source systems are often implemented using flat files and specialised retrieval packages. Most of the integration is link driven and is achieved by the creation of cross-reference indexes. For instance, the SRS language defines search in indexes of databanks, and combinations of queries using and, or and andnot. With respect to the combination of databanks the link construct is introduced. This allows for queries of the forms find all entries in databank A that are referenced in databank B, and find all entries in databank A that reference entries in databank B. The advantage of such systems is that queries relating to knowledge in different databanks can be asked and that the query processing is fast.

However, although this is a first step in integrating data sources, this solution does not handle the differences in terminology used in the underlying sources, is syntax based and only allows limited query functionality over multiple databanks. Also, adding a new resource requires cross-referencing with the other resources. The BioKleisli, K2, TINet, P/FDM, TAMBIS and BioTRIFU systems use an approach based on view integration. Also IBM's DiscoveryLink can be placed into this category. In this approach the underlying schemas are integrated to form a global schema. The global schema is queried in a high-level language such as CPL (e.g. BioKleisli) or OQL (e.g. K2).

The languages in TAMBIS and BioTRIFU have been inspired by the study of the use of current biological databanks. The other proposals seem to be based mainly on database technology. In general, the view approach allows for more complex querying and allows for support for integration on schema level. The advantages of such systems include the possibility of complex querying, the knowledge that is required of the end-user is not as large and the local conceptual models are used in the integration. Further, these view integration systems may also be used to create warehouses. As a possible solution of the terminology discrepancy problem in the local schemas as well as a step towards semantic querying, ontologies could be used.

Ontologies

In bioinformatics, the majority of data sources are databases and tools with a variety of user interfaces. The ontologies help manage the interoperation between these resources by providing a dynamic controlled vocabulary of concepts. One widely used definition of ontology is "a specification of a conceptualisation". Even with this definition, there is a spectrum of the structure of ontologies, ranging from at lists of controlled vocabulary to very formal logic-based implementations.

Pragmatically, an ontology typically includes a hierarchically arranged list of concepts (i.e., classes) of a given domain, relationships among these concepts, definitions of these concepts and relationships, and optional logical axioms that serve as further constraints among these entities. Apart from the high-level aim of representing the knowledge of a domain in a computationally amenable form, an ontology provides a controlled vocabulary in that a given term always has the same meaning. Thus, rather than attempting to parse the complex structure of natural language for embedded biological knowledge, computational agents can query an ontology and a knowledge base based on the ontology to specifically and reliably retrieve data.

BIOINFORMATICS RULES

A natural extension of factual knowledge are rules and logic programming. Here we have three application areas in which rules are used: Rules to represent metabolic pathways and query them in computation tree logic; Rules as constraints for structure prediction; and rules learned by inductive logic programming to declaratively capture biological knowledge.

Uses of Computation Tree Logic

This idea of introducing formal methods was mentioned as a prime motivation by early efforts at formal modelling, and the specific prospects of using computational logics were clearly articulated in. Substance was given to this idea by proposing the use of the Computation Tree Logic CTL as a query language for biomolecular networks. In networks biology, the complexity of the systems at hand (metabolic networks, extracellular and intracellular networks, networks of gene regulation) clearly shows the necessity of software tools for reasoning globally about biological systems.

Several formalisms have been proposed in recent years for modelling biochemical processes either qualitatively or quantitatively. State-of-the-art tools integrate a graphical user interface and a simulator, yet few formal tools are available for reasoning about these processes and proving properties of them. The focus in Biocham has been on the design of a biochemical rule language and a query language of the model in temporal logic, that are intended to be used by biologists. Biocham is a language and a programming environment for modelling biochemical systems, making simulations, and querying such models in temporal logic. Biocham is composed of:

1. a rule-based language for modelling biochemical systems,
2. a simple simulator,
3. a powerful query language based on Computation Tree Logic CTL,
4. an interface to the NuSMV model checker for automatically evaluating CTL queries.

Biocham shares several similarities with the Pathway Logic system implemented in Maude. Both systems rely on an algebraic syntax and are rule-based languages. One difference is the use in Biocham of CTL logic which allows us to express a wide variety of biological queries, and the use of a state-of-the-art symbolic model checker for handling the complexity of highly non-deterministic models.

Bioinformatics and Constraint

The protein structure prediction is one of the most important unsolved problems of computational biology. It can be specified as follows: Given a protein by its sequence of amino acids, what is its native structure? NP-completeness of the problem has been proven for many different models. These results strongly suggest that the protein folding problem is NPhard in general. Therefore, it is unlikely that a general, efficient algorithm for solving this problem can be given.

Actually, the situation is even worse, since the general principles why natural proteins fold into a native structure are unknown. This is cumbersome since rational design is commonly viewed to be of paramount importance e.g. for drug design, where one faces the difficulty to design proteins that have a unique and stable native structure. To tackle structure prediction and related problems simplified models have been introduced. These simplified models have been successfully used by several groups in the international contest on automated structure prediction. They are used in hierarchical approaches for protein folding.

Given a protein sequence, then one first enumerates all low (or minimal) energy conformations in a simplified model. In the simplified model, only some aspects of the protein structure are modelled (for which reason they are also called low-resolution or coarse-grained protein models). Then, the say 10 000 best structures are taken and fine-tuned using other methods. These methods usually incorporate biological knowledge and simulation of protein folding on full atomic detail (i.e. molecular dynamics simulation).

Constraint-based approaches to structure prediction express distance constraints on consecutive and adjacent monomers. E.g. to express that the distance between two successive monomers is 1, we introduce for every monomer *i* the three variables $Xdiff_i$, $Ydiff_i$ and $Zdiff_i$. The value range of these variables is [0...1]. Then we can express the unit-vector distance constraint by

$$Xdiff_i =: Abs(X_i - X_{i+1})$$

$$Ydiff_i =:$$

$Abs(Y_i - Y_{i+1})$

$Zdiff_i$ =:

Abs(Zi – Zi+1)

$Sum[Xdiff_i, Ydiff_i, Zdiff_i]$ =:

1

On some models such a constraint-based approach can outperform other existing approaches such as Monte-Carlo simulated annealing, genetic algorithms, purely heuristic methods like hydrophobic zipper and the chain growth algorithm, as well as complete enumeration.

Inductive Logic Programming

Inductive logic programming uses examples and background knowledge to construct hypotheses, which explain the examples using the background knowledge. Muggleton et al. have applied inductive logic programming to a wide variety of biological problems such as the following:

— *Structure-activity prediction*: In it was shown that ILP system Golem was capable of constructing rules which accurately predict the activity of untried drugs. Rules were constructed from examples of drugs with known medicinal activity. The accuracy of the rules was found to be slightly higher than traditional statistical methods. More importantly the easily understandable rules provided insights which were directly comparable to the relevant literature concerning the binding site of dihydrofolate reductase.

— *Mutagenesis*: In ILP system Progol was used to predict the mutagenicity of chemical compounds taken from a previous study in which linear regression had been applied. Progol's predictive accuracy was equivalent to regression on the main set of 188 compounds and significantly higher (85.7% as opposed to 66.7%) on 44 compounds which had been discarded by the previous authors as unpredictable using regression. Progol's single clause solution for the 44 compounds was judged by the domain experts to be a new structural alert for mutagenesis.

— *Pharmacophores*: In a series of "blind tests" in collaboration with the pharmaceutical company Pfizer UK, Progol was shown capable of rediscovering a 3D description of the binding sites (or pharmacophores) of ACE inhibitors (a hypertension drug) and an HIV-protease inhibitor (an anti-AIDS drug).

— *Carcinogenicity*: Progol was entered into a world-wide carcinogenicity prediction competition run by the National Toxicology Program (NTP) in the USA. Progol was trained on around 300 available compounds, and made use of its earlier rules relating to mutagenicity. In the first round of the competition Progol produced the highest predictive accuracy of any automatic system entered.

— *Proteins*: Protein secondary structure prediction. In Golem was applied to one of the hardest open problems in molecular biology. The problem is as follows: given a sequence of amino acid residues, predict the placement of the main three dimensional sub-structures of the protein. The problem is of great interest to pharmaceutical companies involved with drug design. For this reason, over the last 20 years many attempts have been made to apply methods ranging from statistical regression to decision tree and neural net learning to this problem. Published accuracy results for the general prediction problem have ranged between 50 and 60%, very close to majority class prediction rates. In an investigation it was found that the ability to make use of background knowledge from molecular biology, together with the ability to describe structural relations boosted the predictivity for a restricted sub-problem to around 80% on an independently chosen test set.

— *Discovery of fold descriptions*: Protein shape is usually described at various levels of abstraction. At the lower levels each family of proteins contains members with high sequence similarity. At the most abstract level folds describe proteins which have similar overall shapc but are very different at the sequence level. The lack of understanding of shape determination has made protein fold prediction particularly hard. However, it is intriguing that although there are around 300 known folds, around half of all known proteins are members of the 20 most populated folds. Progol was applied to discover rules governing these 20 most populated protein folds. Average in class cross-validated prediction was around 70% and many of the rules were judged to be good characterisations of the fold classes.

REFERENCES

Charles R. Cantor, Cassandra L. Smith, *Genomics: The Science and Technology Behind the Human Genome Project*, Human Genome Project Wiley Interscience, 1999.

Stanley Letovsky (Editor), *Bioinformatics: Databases and Systems*, Kluwer Academic Publishers, 1999.

Stuart Brown. *Bioinformatics: A Guide to Biocomputing and the Internet*, Eaton Publishing Company, 2000.

2

Computational Biology

Computational biology is an interdisciplinary field that applies the techniques of computer science, applied mathematics, and statistics to address problems inspired by biology. Computational biology is the application of computational tools and techniques to (primarily) molecular biology. It is the use of computational techniques to model biological systems at various levels of complexity-atomic, metabolic, cellular, and pathologic. It enables newways of study in life sciences, allowing analytic and predictive methodologies that support and enhance laboratory work. It is a multidisciplinary area of study that combines biology, computer science, and statistics. Computational biology is also called bioinformatics, although many practitioners define bioinformatics somewhat more narrowly by restricting the field to molecular biology only.

In computational biology, there are different types of problems such as how to find genes in DNA sequences, to predict the structure and functions of newly discovered proteins, to cluster proteins into families, and to align similar proteins, and to generate phylogenetic trees to examine evolutionary relationships. Therefore, there are different models of genetic change such as the long-term model that deals with the evolutionary changes among species. Another model deals with reconstructing evolutionary trees from sequences, a short-term model that deals with the genetic variations in a population and a final model that deals with finding genes by linkage and association.

Computational biology is one of the most rapidly growing areas of scientific computation. It is driven by two factors, the first of which is the recent explosion in the amount of available biological (particularly molecular biological) data, whereas the second is the commercial incentive to exploit the available biological data for drug development and other discoveries. We see, therefore, that the biological data explosion is due to the increased interest in the study of molecular biology over the past few decades, to the significant improvements in the reliability and speed of automated techniques available to gather molecular biological data over the past fewyears, and to the mapping of the human genome completed recently by both publicly and privately funded efforts (The International Human Genome Project and Celera Genomics Group). These advances have caused the amount of information available in international DNAdatabases to increase exponentially over the last fewyears. Organisation, retrieval, and interpretation of this vast amount of data have become a key computational challenge.

The computational biology field focuses on three key areas of computational molecular biology:

Sequence Analysis: Analysing sequences of DNA and proteins to try to determine their structure, function, and control information. Given two sequences of characters (representing DNA or proteins), the key problem of sequence analysis is to align them to achieve the optimal match between them on a character-by-character basis. The optimality of the alignment is determined by a scoring system that gives credit for suitably matched characters and penalises mismatches. However, gaps may be inserted between characters to try to increase the optimality of the match and therefore may lead to certain complications such as the scoring system penalising gap insertion.

The advantage of aligning sequences is that, if a sequence with unknown properties aligns with a sequence with known properties, it may be possible to extrapolate the structure and function of the unknown sequence. Besides, two sequences with common ancestry may add or remove some amino acids during evolution. There are different types of sequence analysis: pairwise alignment, heuristic pairwise alignment, and multiple alignments.

Structure Analysis: Analysing biological structures to try to determine their sequence and function, and to control information. The key problem in structure analysis is to find an optimal correspondence between the arrangements of atoms in two molecular structures (say A and B) to align them in three dimension. The optimality of the alignment is determined by using a root-mean-square measure of the distances between corresponding atoms in the two molecules.

However, it is not known a priori which atom in molecule B corresponds to a given atom in molecule A (the two molecules may not even have the same number of atoms). Structure is believed to be more closely related to function of proteins than sequence. Structural alignment makes clear the common ancestry of two or more proteins (if such a common ancestry exists). It allows identification of common substructures of interest and it allows classification of proteins based on structural similarities.

Function Prediction: Understanding how sequences and structures lead to the specific functions. The key problem is to predict the function of protein structures based on sequence and structure information. The function is loosely defined, and can be thought of at many levels atomic or molecular. Currently, relatively little progress has been made in function prediction, particularly for higher-order processes. The current methods of function prediction are:

— *Experimentation*: It is used to determine the function of proteins and other structures but expensive in terms of time and money.
— *Annotation Transfer*: When sequence or structure analysis yields correspondences between structures, the known properties and function of one is used to extrapolate the properties and function of the other. This method has been extremely successful, but the facts that similar sequence or structure does not always imply similar function and that the annotated information about the "known" protein or its sequence or structure information in the database may be incomplete or incorrect are drawbacks to be considered.

POTENTIALS IN COMPUTATIONAL BIOLOGY

Following are the explanations of some technical terms that have proved to be essential for researchers in computational biology.

Chromosome

A chromosome is a long string of double-stranded deoxyribonucleic acid (DNA), the molecule that serves as a primary repository of genetic information. Thomas Hunt Morgan found that genes on a chromosome have a remarkable statistical property, that is, genes appear as being linearly arranged along the chromosome and also that chromosomes can recombine and exchange genetic material. A gene is a unit of heredity used to describe a unit of phenotype variation.

Allele

Alleles are alternate forms of the same gene. There may be hundreds of alleles for a particular gene, but usually only one or a few are common. A homologous pair of chromosomes contain two alleles, one in the chromosome derived from the father and the other in the chromosome derived from the mother. If, for example, the chromosome inherited from the mother has a mutant allele at a specific position, this position on a chromosome is called a locus, and the presence of a single mutant allele creates the trait of disease. However, the child will not suffer from the disease caused by this mutation unless both the genes inherited from parents are defective or one of them is on the X chromosome, for example, hemophilus. In brief, an allele is a type of the DNA at a particular locus on a particular chromosome.

Recombination

Recombination or crossing over is defined as the recombination of maternal chromosome pairs with its paternal chromosome and exchanges material in the genesis of a sperm or egg. This formation of new gene combination is the result of the physical event of crossing over. The intensity of linkage of two genes can be measured by the frequency of the recombinants.

The probability that a recombination event occurs between two loci is a function of the distance between these loci. In fact, the alleles at two loci that are far apart on a chromosome are more likely to combine than the alleles that are close together on a chromosome. Genes that tend to stay together during recombination are called linked. Sometimes, one gene in a linked pair serves as a marker that can be used by geneticists to infer the presence of the other genes causing disease.

Meiosis

Before explaining meiosis, let us explain the relationship between genes and alleles. In Figure 1, we notice two gametes inherited from the father AD, which are called the gene 1, and the two gametes inherited from the mother ad, which are called gene 2. Therefore, the formation of haploid germ cells from diploid parent cell is called meiosis.

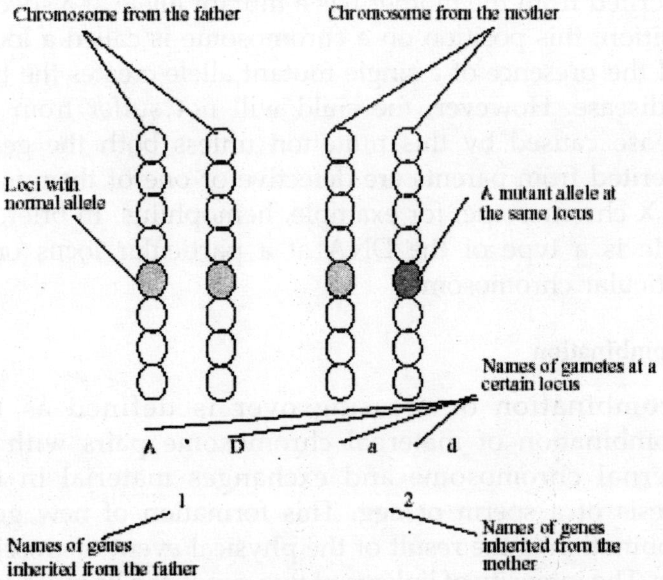

Figure 1. Genes and alleles at a specific locus.

Meiosis is informative for linkage when we identify whether the gamete is recombinant.

Linkage Analysis

Geneticists seek to locate genes for disorder traits (gene disease) among the genome, which is pairs of 23 human chromosomes. The statistical procedure used to trace the transmission of a disordered allele within a family is called linkage analysis. This analysis is based on genes, whose locations on a particular chromosome are already known, and are called markers. Genes will be inherited together if they are close on the same chromosome because recombination is less likely.

Recombinant chromosomes will occur less frequently (less than half of the time) than nonrecombinant chromosomes (more than half of the time). The recombination is measured by the recombination fraction denoted by θ, which is the probability of a recombination between two loci. The lowest possible value for θ is zero, which means that there is no recombination between the marker and the trait locus. In fact, either the marker is the trait or the two loci are so close together that they rarely recombine.

The upper limit of θ is 0.5, which means that the loci are not linked. The recombination frequency θ for unlinked loci is 0.5. In brief, $0 \leq \theta \leq 0.5$. The family is said to be idealised pedigree if no recombination has taken place. If there are recombinations, it is quite easy to calculate θ; it is the summation of the number of recombinants among offspring divided by the total number of offspring. For example, if one recombinant out of nine offspring means that θ is equal to $1/9 = 0.11$.

Frequency of offspring

The computation of expected frequency of offspring genotypes in linkage is as follows: As an example, consider a father who has haplotype AD/ad and a mother with haplotype ad/ad. All the mother's gametes will be genotyped ad. Thus, the probability that the mother gives the alleles ad is equal to 1. Father, however, may have one of the four different gametes, AD, Ad, aD, ad. In addition, the probability that the father gives the allelesAD is equal to the probability that the father gives allele A at marker multiplied by the probability of having no recombination between marker and trait.

In fact, it is equal to $1/2(1 - \theta)$. Similarly, the probability that the father gives the alleles Ad is equal to the probability that the father gives alleleA at marker multiplied by the probability of having recombination between marker and trait. In fact, it is equal to $1/2\theta$. The probability that the father gives the alleles aD is equal to the probability that the father gives allele a at marker multiplied by the probability of having no recombination between marker and trait. In fact, it is equal to $1/2\theta$. Finally, the probability that the father gives the alleles ad is equal to the probability that the father gives allele a at marker multiplied by the probability of having recombination between marker and trait. In fact, it is equal to $1/2(1 - \theta)$. Then, the expected frequency among the offspring is a function of θ.

Multipoint Linkage

There are two types of statistical techniques used in the linkage analysis, parametric linkage analysis and nonparametric linkage analysis.

Parametric linkage analysis uses statistical procedures to estimate θ and sometimes other quantities. The odds for linkage is a quantity that is equal to the ratio of two probabilities; the numerator is the probability of observing the data given that θ is less than 0.5 (i.e., the marker and the trait loci are linked) and the denominator is the probability of observing the data given that θ is equal to 0.5 (i.e., the marker and the trait loci are not linked). The common logarithm (base 10) of the odds (likelihood) of linkage is specific to geneticists for the computation of parametric linkage analysis. It is called the LOD scores.

The second method used in linkage analysis is suitable for complex gene disorders, unlike the first one suitable for single gene analysis. It is called the nonparametric approach. The advantages of nonparametric techniques are that it is not necessary to make assumptions about the mode of inheritance for the disorder; their disadvantage is they are less powerful than the parametric techniques. An example of nonparametric techniques is the affected sib-pair method. The geneticists gather data on a large number of sibships to locate those that have at least two members of a sibship who are affected with the disorder.

The affected sib pairs are then genotyped at the marker locus, and the sib pairs are placed into one of the two mutually exclusive categories based on their genotypes at the marker. The first category includes all sib pairs who have the same genotype at the marker, these being called marker-concordant pairs. The second category is for the marker-discordant pairs, those sib pairs who have different genotypes at the marker. If the marker is not linked to the gene for the disorder, then we should expect an equal number in both categories. However, if the marker is linked to the disease locus, then there should be more marker-concordant pairs than marker-discordant pairs.

Sib Pair Analysis

The sib pair analysis is the probability of having 0, 1 or 2 common alleles. This analysis is known as identity by descent (IBD). When IBD information is incomplete, partial information extracted from marker data may be summarised by the inheritance distribution, a conditional probability distribution over the possible inheritance vectors at the marker locus.

Joint genotype probabilities

The joint genotypes probabilities are conditioned on the number of pairs of genes (one from each individual), which are IBD. The value p_i represents the population frequency of allele *i*. Genes not IBD are assumed to be statically independent. Genotypes are considered to be unordered pairs of genes. For example, consider two offspring with respectively ab and cd alleles; the probability of having one common identical alleles is equal to zero, and the probability of having two common identical alleles is also equal to zero, but the probability of having zero common identical alleles is equal to the multiplication of probabilities of alleles frequency in such a population multiplied by the number of possible states $S(p_a * p_b * p_c * p_d * S)$.

Example of Computation: Consider $A = (1,..., v)$ a set of different alleles. $P_a = (p_1,..., p_v)$ a set of probabilities of alleles frequency. $G = (x_1, x_2, x_3, x_4)$ an ordered vector of alleles from the set A of two individual X with alleles x_1 and x_2 and the individual Y with alleles x_3 and x_4, and s is a random gene state of G belonging to a set S of IBD states.

If s = abcd implies that there is zero IBD ($m = 0$). If s = abbc implies that there is one IBD (x_2 and x_3 are common). Assumed that x_1 and x_2 are never IBD and also x_3 and x_4 are never IBD. The technical definition of this assumption is that the individuals X and Y are not inbred. This implies that the set of states S contains only IBD states compatible with this assumption and the resulting states are mutually exclusive.

Joint distribution of genotype

The distribution of a set of genes is derived by enumerating all possible IBD states, then calculating the probability of each of the states. These probabilities depend only on the pedigree itself, while the actual probability of observing a given genotype is calculated by conditioning on the IBD states, recognising that non-IBD genes are conditionally independent.

Pedigree probabilities

For a pedigree composed of k persons, each person K has an ordered genotype relative to many locations (called loci): $X_k = (x^1_k, x^2_k,..., x^l_k) = (x^1_{km}, x^1_{kp}, x^2_{km}, x^2_{kp},..., x^l_{km}, x^l_{kp})$ where x^i_{km} is the allele inherited from the mother at locus i and x^i_{km} is the allele inherited from the father at locus i. The problem is to calculate the probability of phenotypes (unknown genotype) of all persons of a family, when knowing the genotype at different positions.

Another way of defining the problem, given the family with many loci in an order and the phenotypes of members of the pedigree, given also the recombination's fractions between consecutive loci θ_i, the problem will be to reconstruct the best genetic map by maximising the chance of having data presented and to replace the old recombination fractions by the new one calculated by the algorithm. Genetic mapping is placing genes and other genetic markers on chromosomes, ordering them in relation to one another and other landmarks (such as centromeres), and assigning genetic distance to adjacent pairs.

There are three types of probabilities:

1. *Founder Probabilities*: If the person K is a founder, the probability of the ordered genotype across all loci is equal to the multiplication of all allele frequencies at different loci. $P[X_k] = P[x^1_k] * P[x^2_k] * \cdots * P[x^l_k]$. If X_k has as alleles 1 and 3 at locus l®$P[X^l_k = 1, 3] = P^l_1 * P^l_3$. Thus, the allele frequency at locus l is the multiplication of population frequency of each allele at the specific locus.

2. *Transmission Probabilities*: This is the probability that a child inherits a particular genotype from his/her parents. $P[X_k = x_k \mid X_m = x_m, X_p = x_p]$, then all genotypes are ordered $P[X_k \mid x_m, x_p] = P[X_{km} \mid x_m] * P[X_{kp} \mid x_p]$.

3. *Penetrance Probabilities*: This is the probability of phenotypes. When genotypes are given, they are compared with phenotypes; if they are compatible then the penetrance probability is equal to 1, otherwise it is equal to zero.

Performing calculations on multi-locus data means calculating probabilities on pedigrees using different algorithms. The traditional algorithm is to choose new θ values by approximating the derivative of the likelihood function $L(\theta^{old})$. The main disadvantage of this method, called quasi-Newton, is that the new θ may have lower likelihood. The Expectation and Maximisation (EM) algorithm overpasses this drawback by offering a powerful general approach to obtain maximum likelihood estimates from incomplete data. The algorithm starts by making an initial guess of θ distribution. Then, it calculates the expected values for complete data such as the number of recombinant and nonrecombinant meioses in each interval. Afterward, it maximises the likelihood estimate θ^{new}.

The algorithm keeps on repeating expectations and maximisations until the likelihood converges to a maximum. The theoretical advantages of EMsearch are: less computation time per iteration, increased likelihood on each iteration, good initial convergence properties, exact expressions for derivatives of the likelihood, and ease of generalisation. However, the problem is also stated as the construction of multi-locus linkage maps in humans or as a missing data problem. The efficient solutions are to determine the expected number of recombinations that occur in each interval, given the recombination fractions $\theta = (\theta_1, \theta_2, ..., \theta_l)$ and the phenotypes for all loci $(l_1, ..., l_l)$.

A special case reconstruction: in this case we suppose that we can observe the genotypes of each individual in a pedigree, including which alleles are on the paternally and maternally derived chromosomes. In each interval, it is then easy to find out by inspection if there is recombination or nonrecombination. Thus, the computational time is proportional to the square of the number of loci. Elston-Stewart algorithm: this is the general case of genetic reconstruction, the algorithm is applied recursively up the family tree, with probabilities computed for each possible genotype of each child, conditional on the genotypes of his parents, the phenotype of the child, and the phenotypes for the child's descendents. The computation time scales with the number of alleles on each locus. This becomes impractical if the number of loci is more than 5.

Hidden Markov algorithm: this algorithm can be applied with any number of loci but with pedigrees of limited size. It is based on a nonhomogeneous Markov chain of inheritance vectors $v_1,..., v_l$ at different loci with known transition matrices between any two consecutive loci. An inheritance vector vi is a vector of 2* nonfounder binary bits. A coordinate is 0 if the allele is inherited from the father; otherwise, it is 1. Thus, the molecular biological approach to understanding the flow and expression of genetic information involves studying the structure of macromolecules (DNA, RNA, and protein) and the metabolic steps that mediate the flow of information from the genome to the phenotype of the organism. With the advent of large databases of genetic information for these macromolecules, a completely new approach to studying gene expression and its regulation has developed. This field is known as bioinformatics.

MAJOR PARADIGMS IN CB

In some ways, CB-especially bioinformatics-is called upon to clean up the mess created by the recent collision of three major paradigms:

— robotics,

— miniaturisation,

— parallelisation.

Robotics allows efficient high throughput repetition of simple tasks. Miniaturisation of biology enables high experimental densities on ever-smaller amounts of material. Parallelisation is a thought process that says, "Why not do multiple experiments at once, just like computers process many data streams at once?" The result is high density experimentation resulting in the generation of so much data that humans are no longer able to comprehend the information hidden within, convert it into knowledge, and take appropriate action without the aid of computers. Furthermore, the globalisation of research, either by multinational pharmaceutical companies, or academic collaborative networks, would be impossible without computers. Both data and compute services must be available anytime and anywhere to global research networks, or industries.

Innumerable biotechnology companies, as well as major pharmaceutical companies and academic institutions have invested heavily in CB. A dynamic market with hundreds of suppliers, thousands of practitioners, and a myriad of approaches has evolved. There are about 1273 biotech companies in the US with total revenues of about $22B, and R&D expenditures of about $11B. They employ approximately 162K people. There are about 1570 biotech companies in Europe with total revenues of about $8B, R&D of about $5B, and employing 61K people.

Recently, it was reported that the year 2000 was a record year in venture capital investments; some $33B was pumped into biotechnology globally. IDC expects the bioscience IT market to grow from about $2.2B in 2000 to $11.6B by 2004, a compound annual growth rate of 52%. Storage will represent nearly 43% of the revenue by 2004, servers will be 34% and services 22%. There are many federal and state programs in support of bioinformatics and post-genomics activities (the NIH is obvious, but the Department of Energy actually sparked the Human Genome Project, and still supports the field).

Most every state and federal government wants to have one of the "top 20" bioinformatics sites in the world located in their state or country. Recently the European Commission announced the grant of 19.4M euros to support new generation bioinformatics. The project involves 25 labs in 11 countries, coordinated by the EBI

(Hinxton, UK; outstation of the EMBL, Heidelberg). Canada has also recently pledged unprecedented support at both the Federal and Provincial levels for CB research. Patenting of human genes (Table 1) has raised a lot of ire since people are mistakenly concerned about "some company owning my genes."

Patents, however, only deal with the commercial use of a discovery and can encourage researchers to do the research to find those uses. Employment opportunities abound in CB, and salaries have been bid upwards, especially in bioinformatics over the past few years (but may be hitting a plateau now). For example, 2 out of 3 of Celera's employees are software engineers (Celera is Craig Ventner's company that published a version of the human genome in competition with the public Human Genome Project). A number of Universities are starting to offer concentrations in bioinformatics, e.g., University of Manchester (U.K.), among several, has a well-known web-based program.

Table 1. Patenting of Genes

Company	Number of Patents (approx)
Human Genome Sciences	7500
Incyte	7000
Hyseq	5500
Millennium	500

DNA is the biological equivalent of the Jacquard Loom (one of the conceptual origins of the electronic computer). The loom used a paper tape to control a mechanical loom that wove patterns in fabric. The paper tape controlled the product, but it didn't say anything about how to staff and run a textile business, how to get appropriate supplies to the factory in time, why the textile industry declined over time, etc. DNA (which includes sub-sequences that form genes), through several biochemical steps (transcription into messenger RNA through to translation into protein), achieves its "purpose" by storing the code for producing proteins, the real workhorses of biology.

Genomics is the study of the entire gene sequence of an organism. Genes are "expressed" (turned-on) or turned "off" at different times in response to different biochemical or environmental perturbations. The expression of genes (or the lack thereof) gives rise to differentiated organs during ontogeny (the history and growth of an individual organism from conception) and contributes to the susceptibility to, and/or onset of, disease. Different cells also produce different proteins at different times and/or in response to environmental factors (drugs, disease, allergens, etc.). The entire protein output over time is called "proteomics". The majority of drugs are designed to interact with proteins, but some drugs target DNA, RNA, membranes, etc.

The life science community is not only drowning in data, but new data types and formats are emerging daily. The problem is further exacerbated by a lack of a simple, controlled vocabulary. Recently, it has become fashionable to talk of cellomics, metabolomics, toxicogenomics, and other "...omics", all meaning the parallel, high throughput determination of all (or most) of the cellular activities, metabolites, toxic effects, etc., using micro- and nano-array, and other, technologies.

Currently about 1000 genomes have been studied at some level of detail (bacteria, viruses, insects, small plants, mammals, etc.), and some proteomes are now being constructed in the dawn of this post-genomics era. The fields of proteomics and systems biology are expected to advance as rapidly as their predecessor-genomics-once approaches and methodologies are first standardised, then scaled for higher throughput and made broadly accessible to the whole community and not restricted to an affluent industrial subset of the community. Genomics provides biological targets (receptors, enzymes, etc.), as well as models for drugs through proteomics.

For example, once a protein or peptide is identified as a good candidate, medicinal chemists can then use the protein or peptide structure (structural genomics) to help guide the design of small molecules, called peptidomimetics, that mimic the key features, but are more stable, more easily formulated, and cheaper and easier to produce. Proteins/peptides, as drugs per se, have many problems in delivery and formulation, so there is continued interest in small molecule drugs. The biochemistry of life is compartmentalised by cell and organ or system type. The functioning of an entire living organism is both fascinating and the object of intense study involving many disciplines.

COMPUTATIONAL BIOLOGY IN MEDICAL FIELD

Many of the points have close parallels in agriculture and industrial applications. Pharmaceutical development faces severe hurdles in clinical trials, whereas agricultural products must face environmental impact, global distribution in uncontrolled environments, cross-hybridisation risks, conservative farming practices, and, in the case of genetically modified foods, consumerism. The pharmaceutical industry is the most profitable sector of the Fortune 500 (profits of 18.6% of revenue, compared to median of 5% for all industries), with the highest return on assets and shareholder equity, as well.

However, there are many problems, e.g. the cost of research, the short effective patent life due to long approval process, intense competition, increasing sales and marketing expenses, pricing pressures, consumer and governmental backlash, and so forth. It has been calculated that big pharmaceutical companies (e.g. Pfizer) needs about 3-5 New Chemical Entities (NCEs) per year to maintain double-digit growth rates in the long term. However, R&D

costs have been rising, while the NCE's have actually been dropping, over the last decade to as low as 0.6 NCEs per pharma per year. At the same time, recalls have been increasing.

With typical development times of 10-15 years and costs of $300-500M, there is intense pressure to find newer discovery and development paradigms. However, one should remember that most of this money is spent on clinical trials and not on basic research. Therefore, the selection of compounds for clinical trials has to be improved since pre-clinical development represents only about 40% of the R&D spend; only the most likely to succeed show go into clinical development. In the late 1800's both schizophrenia and syphilitic insanity were treated as the same "disease" based on common symptomatology.

Medical practice, in fact, codified and classified diseases based on observable symptoms. It is only relatively recently that biochemical, pathological and, now, genetic diagnostics are starting to deconstruct this picture of disease and replace it with more objective criteria. Single diseases are now being subdivided into sub-classes, just as biochemical receptors were subdivided over the last 20 years. Biological Discovery is now all about finding "validated" or "druggable" molecular targets, whenever possible, and using these to discover novel therapies. In biopharma, a "biologist" is typically a biochemist, pharmacologist, cell biologist, or physiologist, as well as molecular biologist or geneticist.

A "target" is typically a protein in the form of an enzyme that performs some chemistry, or receptor that binds a molecule or undergoes some transient structural change and releases some other messenger molecule that then binds to another protein. Ultimately there is an observable response: lowered blood pressure, control over cell growth, etc. Sometimes, DNA, RNA, or membranes are targets. Just because you can link a gene to a disease doesn't mean that you have found a good drug target in the gene or its protein products.

Inhibiting a particular target might actually make things worse because of biochemical feedback mechanisms, or because the target is somewhat similar to another and gives rise to a lack of selectivity, or it is sequestered and inaccessible. Until recently, all drugs worked on only about 300-500 known targets under the misconception that a single gene corresponded to a single protein. This fundamental tenet of biology was shattered in 2000 upon completion of the human genome project where it is numerically implausible that 30,000 genes could account for the diversity seen in mankind and the variety in the proteins (structure, composition, and timing of appearance) documented over the last decade. The explanation is thought to lie in epigenetics (methylation and other transformations of DNA and RNAs to turn on/off genes), alternative splicing mechanisms, proteomics, and other phenomenon that accentuate the diversity.

Genomics will increase the number of potential targets-to perhaps. 1000 to 5000-but each gene (or, its protein products) is not going to work well as a "druggable" target. Some believe that each gene encodes up to 10 proteins, which greatly increases the number of potential targets available for study. Whatever the number, only a percentage will ultimately work as useful drug targets.

Chemical discovery often means the elucidation of "small molecule" drugs, typically under about 400 molecular weight, and typically "organic" (comprised of carbon, hydrogen, and some oxygen, nitrogen, or sulphur, plus other elements). These types of compounds are easier to manufacture and formulate then larger molecules. However, proteins, peptides, or larger molecules can be used (e.g., antibodies or modified viruses). Such biotechnology products can do very well in the marketplace. Since 1982, FDA approved 76 biotech drugs and another 369 are in clinical trials.

Chemical discovery has been greatly helped by robotised techniques called "Combinatorial Chemistry," which produce "libraries" of dozens to millions of compounds by various techniques. Compounds are synthesised (by hand or combinatorial chemistry) by medicinal chemists, tested by analytical chemists, and scaled up by process chemists. Computational chemists help in the design, optimisation, and selection of compounds so that the fewest number of compounds will encompass the greatest number of chemical functional groups and possible interactions, minimum side-effects, and optimised physical properties.

Pre-clinical development involves studying selected candidates for in vivo adsorption, distribution, metabolism, elimination, and toxicity properties (ADMET), doing synthetic scale-up (from picograms to milligrams, or milligrams to kilograms), setting up analytical procedures, and discovering how to formulate the molecules into pills, injectables, etc. Before the advent of genomics, miniaturisation and laboratory robots, expensive ADMET studies were almost always done on the relatively few compounds selected for pre-clinical development.

The trend now is to use in vitro and computational screening techniques early in the discovery phase to increase the chances that candidates eventually selected for development will survive the more extensive and expensive animal studies of a full development program. This is now called "fail early" strategy, adopted by several companies such as Roche, and has resulted in re-design of the entire discovery process. Roche reports that they now go from target to clinical candidate selection in as little as 22 months.

Clinical trials require a series of well-defined studies. Phase I is conducted in a small group of healthy volunteers to check for absorption/excretion, maximum tolerated doses, side effects, etc. About 61% of the candidates fail Phase I. Phases II-III involve larger groups of sick patients in an attempt to prove safety and efficacy, either alone (Phase II), or in combinations (Phase III). About 50% and 26% of the candidates fail Phase II and III, respectively. Phase IV consists of post-marketing surveillance for adverse effects that might occur in small percentages and therefore be undetected by the clinical trials (which may involve only a few hundred patients).

Overall, only about 4% of drug candidates entering preclinical development make it to the NDA (New Drug Application). However, about 60% of the R&D budget is spent on clinical trials and development. Commercial production may involve making milligrams to kilotons of materials to very high (GLP, GMP, ISO 9000/1) standards, and occurs after candidates show promise in expensive and time-consuming Clinical Trials. A company may also spend tens of millions of dollars in setting up commercial production facilities on the gamble that the candidate will be eventually approved by the regulatory agencies. Or, they may contract with Contract Research Organisations (CROs; or Contract Service Organisations CSOs) or Contract Manufacturing Organisations (SMOs) to perform varying amounts of research, development, and/or manufacturing.

We can divide CB into several areas. Bioinformatics deals with the information processing requirements of genomics and postgenomics activities, viz. DNA sequence assembly, gene finding, and analysis, as well as the follow-on "post-genomics" steps of gene and protein expression and regulation, up through comparative genomics and pharmacogenomics. BioSimulation is the computerised simulation of biological processes, which includes BioNetworks ("Systems Biology") and/or Metabolomics, an attempt to put much (eventually all) of the biological networks (like Krebs cycle) in the computer to simulate biology. This is very useful for figuring out where the good drug targets are, for example, and is emerging as an area of active interest and development.

The complexity that will be involved is mind-boggling as the analyses that will be required to examine the protein and gene expression patterns of whole systems (cells, tissues, organs, or even organisms) in relation to each other within the context of the experimental parameters (disease state, or time course, etc.) are very complex to simulate. Macromolecular structure is the study of the 3D structure of large molecules like peptides, proteins, antibodies, and the interaction with other molecules. "Structure-based drug design" includes all of the techniques for "rational" drug design of the 1970s-80s, plus docking, artificial intelligence, and other techniques for computational structure modelling and interaction. Most of the effort is on proteins, but there are a number of groups looking at RNA and DNA structure, as well.

Clinical informatics (biomedical informatics) refers to the tracking and analysis of patient clinical records. Some people are starting to look at medical records and relating them to genomic information. Although fairly well established, clinical informatics holds the greatest potential of all the informatics fields to truly impact the cost of drug discovery. Given that roughly 2/3 of the $500M required to identify, test, produce, and launch a new drug is consumed during the clinical trials, improved methods for collecting the patient data directly from the physicians and distributing that data securely will have a dramatic effect on the expense of this stage of the drug discovery pipeline.

Chemoinformatics (also called "cheminformatics") refers to the analysis of small molecules, compounds with molecular weight under about 300 daltons. The compounds are combinations of atoms in particular 3-D shapes, generally organic and containing carbon, that can bind to protein, DNA, or RNA targets that have a certain electrostatic charge, affinity for water, and potential for toxicity to a system once absorbed. Since small molecules are often screened against protein databases to find fits (discovery pharmacophores), there is an interface between chemo- and bioinformatics. There is also an interface with the vast stores of non-genomic biological/pharmacological testing data found in every biopharma (related to toxicology, absorption, and adsorption, etc.).

Molecular modelling, QSAR (Quantitative Structure Analysis Relationships), combinatorial chemistry (production of small molecules in parallel synthesis) are just some of the subfields represented in this category. We are interested in the computational aspects of molecular biology in the area of Bioengineering, e.g. engineering bacteria or tissues for biocatalysis, rather than the mechanical, e.g. artificial limbs. Medical imaging is of interest because it is computationally intensive scientific visualisation and can be related, eventually, to other aspects of CB.

A unifying view of informatics is depicted by the following data flow view: Raw data (like the position of spots on a slab of gel, or the readout from an instrument) needs to be gathered, archived, and analysed into information, such as translating the spots into the sequence of nucleotide bases A,C,G,T comprising sequences of DNA, or the 3D structure of a protein with the relative location of all of its atoms. The information is studied further and combined into knowledge (the location of genes in the sequence, the proteins that come from the genes, the function of the genes and proteins, how they interact, their relationship to each other, etc.), using a more general theory, which eventually leads to action (the simple decision to run another experiment, or the decision to risk $300M to develop the drug candidate). All of these steps are not only facilitated by computers, they are now absolutely impossible without them. In fact, modern biopharma companies are becoming

Information Processing Companies, (but IT is not their core business!). It has been estimated by Viaken that mid-size biotech companies currently have about 3-10Tb (1Tb=10^{12} bytes) of data, large pharma have 10-40Tb, and genomics data providers have 100sTb of data. Many bioinformatics efforts require daily updates and processing of gigabytes of data, requiring T3 connectivity. Celera requires 1.7Tf of compute power and grows its database at the rate of 15-20GB per day. Only companies that master their IT requirements will succeed: they must collect and process terabytes (soon, petabytes, 1Pb=10^{15} bytes) of heterogeneous data into information, then knowledge, and then take action. Eventually they ship the information to the regulatory agencies. Only after approval does the company become a manufacturer of a tangible product (drug, therapy, diagnostic, etc.).

PROBLEMS IN CB

Moor's Law

Moore's Law states that the density of transistors on a chip will double about every 18-24 months. This has held true for four decades, but may be coming to an end as certain physical limits are reached around 2014 (according to the Semiconductor Industry Association). Moreover, actual CPU processing power (e.g., flops) has not kept up to this rate, and may be flattening out somewhat. Using multiple CPUs to process smaller chunks of a problem (parallelism) can overcome some limitations in single CPU performance. However, most algorithms that can scale tend to plateau at about 8 CPUs.

Fortunately, many genomics (text searching) algorithms can achieve much better parallelisation (a class called "embarrassingly parallel") because they can be deconstructed into a large number of independent searches with little message passing between jobs. The final results are then assembled when the independent jobs are completed. Embarrassingly parallel sequence assembly and analysis problems will become relatively less important with post-genomics techniques that extend beyond sequence comparisons into protein structure, statistics, database comparisons, etc. The computer systems used for the computation can either be symmetric multiprocessing (SMP-many CPUs within a single box which share a single memory space), or clusters of systems (which can themselves be either single CPU systems or SMPs).

The systems in a cluster are typically connected together by a commodity network much slower than the processor. Depending on the exact task and algorithm, the computation can either be limited by the computer's ability to read the data ("I/O bound") or by the CPU's ability to process the data ("CPU bound"), or the network's ability to move data (communications bound). It should be clear that reliance on Moore's law to solve the IT challenges of CB will clearly not succeed. We need much more intelligent uses of what we have: better architectures, better interoperability, higher bandwidth, much more clever algorithms.

Genomic Reality

The volume of biological data, on the other hand, is accelerating. Genbank, a public database of DNA, RNA, and protein sequence information, is doubling about every 6 months. As of August 2001, there were 13.5B base pairs in 12.8M DNA sequence in Genbank requiring about 56GB of storage. As of September 2001, EMBL contained 14.4B bases in 12.5M records (largely overlapping with Genbank). These are just two of several dozen CB databases. The growth rate is probably faster than 2X every 6 months. It will only get worse as more and more gene expression and proteomics experiments ramp up.

It is not uncommon for larger labs to generate several GB of data per day. Databases have recently become more specialised and better curated, but redundancies and database asynchrony is on the rise because of the distributed and collaborative style of research. Therefore, database-to-database comparisons are required for analysis and validation, which consume ever more compute cycles and storage, draining resources from more original research. Furthermore, in research, the databases do not just sit there, they get processed and transformed, calculations are run, intermediate results are stored, derivative databases are created, etc. The work product is equally enormous: While the value of information grows non-linearly with the size of the database (similar to Metcalf's Law for the non-linear increase in the value of increasing nodes on a network), the intermediate calculations grow at an even greater rate.

Sequencing the human genome is an enormous accomplishment; however, discovering the genome sequence is just the beginning! DNA carries the recipe for the production of proteins (peptides, enzymes, receptors, antibodies, structural material, etc.) which carry on most of the "life" processes. Genes are turned on and off (gene expression) in response to homeostatic, disease, and environmental effects, but there are many more proteins and these respond much more rapidly to the environment. Proteomics attempts to identify and chart this course and map biological networks ("systems biology"). Moreover, there are individual differences which need to be understood, by sequencing all or part of many individual's DNA, or by the detection of markers (e.g. Single Nucleotide Polymorphisms, SNPs, or more complex markers), called pharmacogenomics.

Biology is inherently non-linear and complex. Many current models in biology are simplified linear approximations of reality, often derived to fit into available computing resources. While computing resources (CPU density, memory, bandidth, etc.) increase regularly, models are often NOT updated to the new hardware, merely adopted or ported. Historically, many different laboratories created databases, which became popular FTP sites (SwissProt, PDB, Genbank, etc.), but little attention was given to the need to assemble such information in, say, a common browser.

To do so, each laboratory would need to write interconversion routines ("wrappers"). But with ever-expanding scope and depth of information, there are now dozens of important databases and interoperability is a major bottleneck. Even within a company, data interchange among legacy systems is often a major problem, especially for recently merged companies. An emerging trend is the public distribution of XML-based generic browsers, which facilitate the viewing of a variety of data types from popular sources.

The number of NCEs (New Chemical Entities) is dropping as R&D costs are increasing. Furthermore, the major pharmaceutical houses are accounting for a decreasing percentage of new drug approvals (from 30% in the '60s to about 18% in the '90s), the slack being taken up by smaller biopharmaceutical companies. One explanation for the decrease in the number of NCEs is that the types of diseases being treated are more complicated, the trials are longer and there is, therefore, a greater chance for side effects to appear.

US biopharmaceutical companies will spend about $33B on R&D this year. And Gartner predicts IT spending by US pharmaceutical companies to increase from $3B in 2000 to more than $5.5B by 2005, a growth rate of 12.8%. Boston Consulting Group estimates that genomics technologies might eventually cut the cost of developing a new drug by as much as 33% (to about $580M from $880M). Based on a

Pfizer drug study, an additional $3.4B revenue could be gained through the application of this approach.

The claim has been made that IT has shortened a 14-year development cycle for new drugs by 4-5 years (at Human Genome Sciences, for example). At modest sales of $100M/year, this amounts to a very substantial savings over costs. Biopharmaceutical companies also struggle with many other issues, such as the integration of legacy databases, the vast diversity of research data types, lack of data standards, buy vs. write issues with research software, lack of audit trails in research computing, return on investment for hardware/software/consulting purchases, and many other issues.

CHALLENGES OF COMPUTATIONAL BIOLOGY

Genomics

Genome bioinformatics consists of using various bioinformatics techniques to assemble genomes and to find genes and other landmarks in the DNA sequence, a surprisingly nontrivial and somewhat contentious undertaking. With so much money invested in sequencing (the Human Genome Project cost $1.4B over the past 10 years, and sold over 1700 large scale sequencers), the machines will be turned loose on other organisms, pharmacogenomics, SNPs, etc. Sequenom, for example, plans to obtain the genome of 15,000 people by 2002, which should provide statistically useful data for disease linkage analysis.

Modern sequencers produce 1000bp reads/sec and can operate nearly 24x7 for days at a time, with continual improvements in technology increasing the throughput. However, it takes about $30-75M, or more, to do a full genome sequence using current technologies (e.g., mouse, rat, zebrafish, pufferfish). There are some 4600-4800 mammals (and about 1.75M total species according to the

Catalogue of Life project), far too many to do every one, even using accelerated technologies! Which organisms get fully sequenced has become a matter of vigorous scientific debate.

The human genome was found to be full of surprises, e.g. the relatively small number of genes originally reported-about 30K instead of the upper limit of 120K postulated earlier. The intergenic wilderness proved to be populated by a motley crew of intriguing genetic characters: active promoters and regulators of gene expression, comatose pseudogenes, descendants of immigrant DNAs (perhaps horizontally transferred from microbes), vagabond sequences, hordes of tandem short-repeats, and great armies of repetitive elements-some with hundreds of thousands of like-uniformed members. However, the controversy continues, and will continue for some time. A recent report sets the number at about 70K genes when transcript and protein databases, as well as comparative genomics techniques, are used.

Sun has worked with many leaders in the genome effort. University of Wisconsin-Madison sequences E.coli 157 and models rRNA folding. The Beijing Genomics Institute assembles the rice genome and is starting on the porcine genome. Both use a Sun Enterprise? 10000 (E10K). The National Center for Genome Resources and Weizmann Institute also use a Sun E10K in genomic research. Incyte's massive LifeSeq Gold genomics database is served from a Sun E10K. Genoscope Sequencing Center in France performed 72 billion BLAST protein/protein comparisons on a 64-CPU (64GB memory) Sun E10K in less than 2 days using Genoscope's EXOFISH gene prediction method.

The project identified between 28K to 34K human genes. Sun and partner Accelrys announced a successful annotation of 70% of the Neisseria meningitidis genome with over 300 new assignments of proteins. A 24 CPU Sun

Enterprise 6500 and Accelrys' GeneAtlas™database were used.

Genomics is one of the older and most well understood areas of CB. Most computation involves large-scale text searching for somewhat ambiguous patterns. The type of problem that is called "embarrassingly parallel," i.e., the job can be broken up into independent sub-jobs and each fed to a separate processor. There is very little communication needed between processors to determine the intermediate results, which can be assembled into the final result by another processor. The computation is integer (character) based and I/O intensive. Sequence assembly and analysis software is comprised of the well-known BLAST, FASTA, HMMER, and many others.

There are two versions of BLAST, NCBI (public domain), and WashU BLAST (not public domain). FASTA has been parallelised for UNIX SMP and clusters. ClustalW is used for multiple sequence alignment and was originally developed at EMBL. It has been parallelised by Sun and it is CPU bound. HMMER is for sequence profile analysis using Hidden Markov Models (HMM) and was written at Washington U by Sean Eddy. It is parallelised for SMP and clusters, is CPU bound, but scales well over multiple CPUs. Wise2 is for alignment of protein sequences or profile HMMs vs. DNA sequences and was written at the Sanger Center by Ewan Birney.

It is parallelised for SMP and is CPU bound. Results are stored in a myriad of databases, both public and private, including GenBank, the largest repository of sequence data. The size of databases is growing very rapidly, such that some production, or routine, processing takes longer and longer, meaning that updates to processed databases takes longer and longer. Databases can be many gigabytes to terabytes in size. Moore's Law tells us that CPU speed cannot bridge the ever-widening computation gap. More intelligent algorithms, or the use of specific hardware (like TimeLogic deCypher™ cards) is required.

BLAST was developed for pair-wise comparisons, not genome vs. genome comparisons, so new algorithms are needed. The controversy over the number of human genes, even with the genome essentially complete, is an illustration of the difficulty of certainty in biology; therefore, there are still opportunities in algorithmic development. TimeLogic deCypher™ PCI card provides very high hardware acceleration for supported algorithms, which include most of the popular algorithms (HMM, Tera-BLAST, Phrap, Clustering and Assembly, BLAST 2, PSI-BLAST, Affine Smith-Waterman, FrameSearch, Reverse Framesearch, ProfileSearch, ProfileScan, ProfileFrameSearch, ClustalW, Symmetric Frame Independent™). For example, performance of up to 2 trillion letter comparisons/second for ungapped BLAST searches have been obtained.

The cards can switch in a few milliseconds between algorithms under software control ("reconfigurable computing"). They are best suited for "production" environments where researchers are not writing new algorithms, but require fast turnaround, repetitive processing of constantly changing databases, etc. Paracel also offers a suite of accelerated applications with a different technology approach. Cluster computing (e.g. Beowulf clusters) have become popular, low cost, flexible tools, but the Mean Time Between Failures (MTBF) can be low, power consumption and maintenance high, scalability is limited to certain algorithms, and throughput bottlenecks are common.

Large clusters present very difficult architectural problems and the hardware components (CPU, interconnects, storage, etc.) must be balanced to give reasonable throughput. With datasets of several gigabytes (3-4 GB is typical of the whole genome), data transfer rates become limiting, as does the performance and reliability of the (NFS) file server. Even with an "older" field such as genomics informatics, it is clear that a similar algorithm, such as BLAST, can be used for multiple problems, and requires different hardware for an optimal solution.

A simple BLAST calculation on a gene or chromosome can be done on a workstation, sequence assembly can be done on clusters, and whole genome-genome comparisons can be run most easily on large SMP machines to minimise overhead from assembling the results. The common genomics algorithms, such as BLAST, we originally derived for simple searches of DNA snippets versus a modest-sized database. A profitable area of research has been in completely rewriting these algorithms for the types of problems that are of current interest such as whole genome-genome comparisons.

For example, the PatternHunter algorithm, written in the Java? development environment, can compare 9B base pairs of triplicate mouse data vs. 3B base pairs of human data in 20 days on a simple PC, an impossibly long calculation using pre-genome-era algorithms. It is also clear that the lack of agreement in such a "simple" task as gene finding (which currently depends heavily upon the particular methodology used) points to the need for substantially better algorithms and validation techniques.

Comparative Genomics

Comparative genomics is used to help identify genes and compare organisms. An underlying assumption is that conserved sequences (common between organisms) are functionally important. What genes are common between well-studied lab organisms, whose gene function is known, and humans? What fundamental differences are there between bacteria and viruses and humans so that selective drug targets can be exploited? What is the phylogenetic relationship between species? An entire genomic sequence will become a datapoint in large in silico experiments, called comparative genomics, population genomics, etc. Frances Collins has stated that "95% of human genes can be found if we have the mouse [genome, because most mammals have similar genes]."

About 45-47 full genomes listed in databases so far, but others think that 10K genomes (out of about 1.75 million known species) will eventually be determined. Knowing the gene sequence alone, will not answer all questions, for example, comparing the worm C. elegans vs. fly D. melanogaster, the fly's genome is twice the size (180M bases vs. 97M bases) but has fewer genes (13K vs. 18.5K). This is attributed to the number of transcription factors (~700 compared to ~500). Roughly speaking, the fly is more complex because of more complex gene expression, not because of more gene coding regions.

Comparative genomics builds on a base of sequence comparison using BLAST-like algorithms, but then uses substantially more statistical analysis to determine linkages. Workstations up to large SMP architectures work well, depending upon the scale of the problem. Fast access to storage is required. Databases are rapidly increasing in size and number and the search for multigenic, or complex, diseases will require extensive statistical comparisons. Obviously, the more genomes that become known, the more complex the comparative analysis becomes. Much attention is focused on SNPs, but this is only a first approximation since most diseases are multigenic, and may involve environmental factors as well. Multigenic diseases will require much more sophisticated statistical analysis and database search technology.

Functional Genomics

Out of the 3.2B base pairs of DNA, only "3-5%" represents genes. Since many experimental and bioinformatics methods give somewhat different results, there is disagreement on the exact number of genes. Figuring out what genes do is even more complicated. The number of disease genes with at least one disease-related mutation that had been discovered by 2000 was about 1112; however, multifactorial diseases, caused by several genes, as well as lifestyle and environmental factors, are far more prevalent.

Functional genomics uses a wide variety of technologies to ascertain gene function: microarrays (Affymetrix and many others), gene knockouts (Lexicon, Exelixis), cellular phenotyping (Aurora), etc. "Function", however, introduces the complexity of space, time, and amount; that is where, when, and how much of a gene or gene product is formed under given circumstances. Gene Expression profiles are most often determined by the use of microarrays. Ten thousand up to 400 thousand pieces of cDNA are bound to a substrate (glass or plastic slide, nylon, etc.) at precise locations, using robots, in nanogram amounts. The researcher collects mRNA from the cells of interest, converts it into labelled cDNA snippets and then washes the sample over the substrate.

Florescence occurs where there is complimentary binding. There are now many variations in microarray technology, pioneered by Stanford and Affymetrix, commercialised by Affymetrix, Genomic Solutions, Mergen, Beckman Coulter, Genetix, and GeneMachines, among others. The biochip market (microarrays, microfluidics) is expected to grow from $531M to $3.3B by 2004 (an incredible CAGR 65% according to Frost & Sullivan); about 66% will be molecular diagnostics, the rest drug discovery and development according to industry sources. That implies a lot of data and very high growth rates in data, as well!

One of the reasons for such high growth is that unlike a DNA sequence determination (which is relatively fixed for an individual, but varies somewhat from individual to individual), there is no single, definitive experiment in

Gene Expression even for a given individual. It is the age-old problem of the biological experiment, where time, space (location), and concentration become very important because the organism is in constant interaction with its environment. Not only is the number of experiments large, image processing of the plates requires lots of storage, followed by multivariate statistical analysis. Therefore, some people have predicted the need for petabytes of storage in a few years.

Although there have been some attempts at standardisation of data formats, there is still no widely accepted single format for gene expression, let alone other functional genomics techniques. Even simple, correct, statistical evaluation of microarray plates is far from universal. Database, and database comparison, problems for microarray data can be run on workstations with moderate to large memory up through large SMP compute servers, depending on the problem, while complex databases are well-served from SMP servers with large associated, or networked, storage. Simply moving very large datasets around a network can be challenging. There are many opportunities to develop algorithmic techniques to analyse and visualise the tremendous amounts of data that is contained in a microarray experiment.

Proteomics

Once DNA is formed at the conception of an organism, it remains relatively static (unless you live near Chernobyl, smoke cigarettes, or routinely climb tall mountains). Yet this same DNA enables cells to differentiate into the various body parts, and evolves into an organism that both matures slowly over time and responds very rapidly to environmental changes. This evolution is accomplished through gene expression, which introduces the concepts of time and space to biology. DNA, through a series of steps, forms (expresses) messenger RNA (mRNA), which in turn, through a series of additional steps, lead to the production of proteins.

Proteins carry out most of the biochemical processes called "life". If a protein is not formed in the right amount at the right time in the right place (cell or part of a cell), then you end up with an annoyance, aberration, impediment, disease, or death. Proteins are used for structural material (muscle), enzymes (that carry out chemistry), receptors (that recognise various messengers and release others, or undergo some physical change), antibodies (recognise "foreign" molecules or organisms), and smaller bits of proteins (peptides) that act as hormones and messengers.

Some proteins are bound to other molecules, including DNA and RNA, forming very large macromolecules with various functions (such as histones which regulate gene expression). Just as genomics is the study of the entire genome of an organism, proteomics is the study of the entire proteome (collection of proteins) of the cell, tissue, or organism. The big problem in proteomics comes in accommodating the large number of proteins (several times the number of genes, perhaps 300K-1M).

One gene can form, on average, 8 alternative splicing variants. Furthermore, proteins contain many active groups to which other molecules (e.g. sugars, phosphates, etc.) are added and subtracted or rearranged in a delicate choreography. And there are many enzymes that cut and rearrange proteins, as well. While the DNA sequence is "fixed", the proteome varies continuously in each of the 200, or so, types of cell, over time. Therefore, there are an infinite number of experiments that can be done! There are two major components to proteomics: protein isolation and identification, and the study of protein interactions, functions, and pathways. The standard method of protein isolation is 2-D gel electrophoresis (a 25-year-old technique), and the standard method of identification is mass spectrometry (which can now detect as little as a femtomole 10-15 mole of material (a mole is the molecular weight of the compound in gram units).

Mass spectrometry can now be used for protein identification, peptide sequencing, as well as identification of post-translational modifications of proteins. The common method of studying protein interactions is called yeast two-hybrid. Many companies are investigating alternatives to these methods as they are all laced with difficulties. For example, many proteins "interact" indirectly, via messengers, and do not actually contact one another (as is required by the yeast two-hybrid technology). One common thread, however, is that they all generate a lot of data. It has been estimated by industry analysts Frost & Sullivan that Proteomics Bioinformatics will account for $2B out of $6B total proteomics spending by 2005, that is, about a third of the total spending.

Oxford Glycosciences has installed a Sun E10K and a 10Tb storage area network (SAN) to help manage its proteomics data. The SAN has 12 fiber-channel arrays, each with 22x36GB drives, partitioned into multiple domains. Caprion Pharmaceuticals, in collaboration with Sun, Oracle, Veritas, and CGI, will develop and operate Caprion's Bioinformatics Platform, CellCarta. The platform will map (in 3D), store, and analyse the protein differences of normal and diseased cells. The enterprise uses Sun Ray™ thin clients, Sun Blade™workstations, Sun Fire™4800, Sun Fire 3800, and Sun Fire 280R servers, and Sun Grid Engine software.

Bionetworks (systems biology) try to construct part (eventually all) of biochemistry in silico. Some companies have formed that provide limited models of specific diseases or systems, e.g. Entelos. The Gepasi/Copasi effort is led by Dr. Pedro Mendes, Virginia Bioinformatics Institute, a Sun Center of Excellence in Bioinformatics. The EcoCyc e.coli pathway database by Peter Karp of SRI runs on a Sun server.

The nature and structure of massive proteomics databases, and fast access to these databases will become a major challenge as researchers struggle to accommodate time, space, and concentration data. Many derivative databases, some requiring extensive precomputation, can and will be created from primary data. Data representation will be an interesting problem in order to handle the space and time and concentration problems of proteomics. Bionetwork simulations run well on large memory machines because of the large number of messages that must be passed during the optimisation of large numbers of equations. Depending on the level of detail desired, bionetwork simulations can require teraflop resources. Large databases and complex algorithms may require grids to achieve computational throughput.

Pharmacogenomics/Single Nucleotide Polymorphisms

It has been observed that about 30% of patients with high blood pressure will suffer adverse consequences, 5-10% will die from stroke, but the rest will likely remain asymptomatic; in 60-74 year olds, Mexican Americans have more than twice the incidence of diabetes as Caucasians. Before the advent of genomics, biological science had a somewhat "blurry" vision of individuals. In drug development, for example, all patients enrolled in a clinical trial looked the same; they were, after-all, estimating the response of a uniform population to their drug. Assuming "uniformity" of the population is a basis of statistics.

But there were always a certain number that responded (hopefully a statistically significant number), a certain number that didn't respond, and, unfortunately, a certain number that may have gotten worse, or had bad reactions or side effects. It was soon recognised that certain ethnic groups might have a particular response pattern, but such observations entered medical practice as lore. Genomics changes all this, making individuals, at least cohorts of individuals, different. Cohorts can now be precisely identified based on their genetic profiles, not just on easily observed characteristics (like ethnicity). It turns out that many of these response differences stem from a group of metabolising enzymes (the Cytochrome P450 system) that exist in many variants, the exact ratio of which depends on genetic (and some environmental) factors.

Polymorphisms in the P450 enzyme 2D6 for example, occur in 7-10% of the population overall (higher or lower in some ethnic groups) and because this liver enzyme metabolises about 25% of all drugs, a deficiency in 2D6 is a good indicator for problematic drug toxicity. Affymetrix has a DNA microarry with all 18 known variants of 2D6 as well as 2C19 variants, but costs too much for general use ($119 for a single use). There are variants that determine responsivity to various drugs as well as adsorption, distribution, metabolism, and elimination properties.

Pharmacogenomics deals with the individual genomic difference in response to disease and drugs based on genomic variance. One example of pharmacogenomics is in the area of childhood acute lymphoblastic leukemia (ALL), which afflicts about 2400 children each year. The compound 6-mercaptopurine 6-MP is effective against ALL; however, about 0.3% of the children have a toxic reaction to 6-MP at standard dose due to an inherited deficiency in thiopurine methyltransferase TPMT metabolising enzyme for 6-MP.

A diagnostic test has been developed which allows screening for the TPMT gene; sensitive individuals can receive smaller doses and the majority receive the appropriately higher dose. GlaxoSmithKline is evaluating a test for its HIV/AIDS drug Ziagen, which can cause severe hypersensitivity reactions in about 5% of patients. The impact of these observations is enormous: from drug discovery to medical practice, it will now be possible to more carefully select drugs for individuals that have activity, selectivity, low side-effects, and toxicity. The expectation is that compliance will be higher. However, the economic decisions to develop drugs with smaller (but more compliant) markets are complex.

The SNP Consortium estimates that within 5 years at least 50-80% of clinical trials will involve genotyping of some sort. A paradigm shift towards individualised medicine versus global blockbuster drugs has begun as pharmaceutical companies are now equipped with the technology to tailor their drugs to a specific target audience, launch quickly, and secure the population as lifetime consumers. This approach drives a tailored model that allows pharmaceutical companies to differentiate their product lines and zoom in on their ideal drug consumer and win their loyalty. How are these cohorts defined? Single Nucleotide Polymorphisms (SNP) is the first level analysis (most diseases are multigenic, not monogenic, with environmental and lifestyle components).

A SNP is a single letter "misspelling" that occurs about every 1000 nucleotides, and at a frequency in a specific population of at least 1%. These are not random genetic mutations. Since there are about 3.2B bases in the human genome, there are about 3.2M SNPs. The large public SNP Consortium has found about 2M, much more than its original target of 300K due to improvements in technology. SNPs occur scattered throughout the genome, but less than 1% fall in protein coding regions (exons) of genes. There may be 10-60K SNPs that actually cause a malformation of a protein (enzyme, peptide, antibody, etc.) by changing the coding sequence of DNA.

Other SNPs can fall in the regions near genes that cause transcription problems (problems in the levels of expressed gene products), but most SNPs occur in non-functional areas of the gene. A cluster of SNPs that "travel together" is called a haplotype, which appear to be very useful in predicting disease. The market for SNP analysis will grow to $750M by 2005, with several components. The use of

SNPs in basic research (using SNP information to choose targets or understand disease process) is already a burgeoning market, clinical diagnostics based on SNPs are developing quickly from a slower start. Whole genome analysis (or whole genome SNP scans) are awaiting much lower costs (<<$.01 per SNP if there are 1M SNPs!) but it may not be necessary to perform such full genome analysis if microarrays (or other technology) can be developed to detect all SNPs.

The revenues in the SNPs market are most likely to be derived from the production of diagnostic kits that will be used to screen individuals for candidacy in drug trials during testing and for drug compatibility once the drug is approved. Databases could become very complex if they involve clinical records as well as genomic information because you might need to link family or cohort members, as well. But cohort members are not readily identified without the genomic data to begin with; therefore, these databases need to be very deep.

For basic research, attempting to control at least some of the variability can be useful; therefore, it has become popular to recruit populations that are relatively homogeneous and have good medical records. DeCODE genetics was one of the first in this area of population genomics, assembling the Icelandic population, and other projects have formed around other groups as well.

Very large, distributed, networked, heterogeneous, and secure databases are expected, which requires excellent database serving, efficient storage, high bandwidth I/O, and web services. Databases will have to integrate molecular, molecular biological, genomic, biochemical, cellular, pharmacologic/pharmacokinetic/ pharmacodynamic, and clinical data, and a myriad of supporting data such as functional, comparative, and structural genomics data. These present some of the biggest challenges because such databases are distributed over widely dispersed geographies. SNP analysis is an intensive statistical analysis, with potentially hundreds to hundreds of thousands of individuals to be compared, and runs well on clusters of SMP.

High Throughput Screening (HTS)

High throughput screening (HTS) resulted from the combination of accurate pick-and-place robotic arms with the miniaturisation of biological assays, which, in turn, was enabled by the development of ever more sensitive analytical technologies (initially fluorescence detection, but others as well) and the idea of doing multiple experiments in parallel. This advance caused two further problems, a) if these robots could screen compounds much faster, then where will the compounds come from how to handle all the data (companies like BMS and Roche can run 1M assays per week from 200K compounds per day, or about 150 full library screens per year)? Frost & Sullivan predicts that the total US HTS market to be $2B in 2000 and could exceed $3.5B by 2004. They predict that the average number of tests per week in the larger pharmaceutical company might approach 7M by 2004.

A medicinal chemist, traditionally, might synthesise a few dozen to a few hundred compounds per year, depending upon complexity and other factors. This rate of synthesis is totally inadequate to "feed the HTS monster." Combinatorial chemistry (CC) is the robotics-aided synthesis "in parallel" of dozens to millions of compounds. Imagine two different rows of chemicals of two different types (e.g., one row of related amines, one row of related carboxylic acids), which when mixed Chinese-menu style, react to form products (amides). If there are 30 amines and 100 acids, then there are 3000 possible combinations (called a "library").

Robots speed up the sample handling and enable the high-speed synthesis and purification of compounds.

Companies such as ArQule, Albany Molecular Research, and Pharmacopeia dominate the commercial production of chemical libraries. Initially, fairly random libraries were tested, but these proved to be low yielding. It is now more common to use computational chemistry techniques of various types to design more focused, "drug-like," libraries, in silico. These so-called "virtual libraries" are screened against various algorithms that try to predict ADMET properties based on structure. In this way entire regions of structure-activity-space-those most likely to be problematic-can be eliminated before actual synthesis begins.

But we also mentioned "millions" of compounds above, how is this accomplished? It is possible to attach chemicals to inert beads and through some clever splitting and pooling of batches of beads to "grow" many combinations of chemicals (peptides are great for this) on the beads. The beads can then be tested in bulk and the active compounds isolated and identified later. There is also a class of combinatorial biology techniques that enable the production of millions of peptides using techniques like phage display. A phage is a virus that infects bacteria. It is possible to manipulate the phage's DNA to express millions of peptides on its surface.

In the past a pharmaceutical company might screen 10-30K compounds per assay, but now 500K is typical; array densities of up to 100K spots/slide can be obtained by several manufacturers. 100K compounds/robot/day "uHTS" is coming online. For example, Pfizer has expanded screening capacity from about 3.8M data points per year to a projected 150M+/year data points by 2002. $3.6B spent worldwide on HTS for lead discovery now, half in US alone, growing at 15% a year. Moving from 96 well plates to 384 (common now), with some running at 1536, and 3456 in research.

The computational problems of computational chemistry go far beyond LIMS (Laboratory Information Management Systems) problems of sample tracking, raw data storage, preparation of reagents, etc. The most valuable libraries are those that have members that might have the highest chances of becoming real drugs. There is a mini-industry involved in the discovery and commercialisation of algorithms and technologies for predicting such properties and designing optimal libraries.

Several companies have formed: Health Designs (bought by Oxford Molecular, later sold back), Camitro (bought by ArQule), Trega (bought by LION Biosciences). Many of the older Computer Aided Drug Design (CADD) techniques (statistical, quantum and molecular mechanical and docking) of the 70s-80s are finding rebirth in the area of ADMET (adsorption, distribution, metabolism, elimination, toxicity) prediction for library design. About $3B per year is currently spent on in vivo ADMET studies and the hope is that in silico studies will be able to capture at least a part of this.

Furthermore, customers are demanding higher and higher levels of purification and sample identification, which raises numerous software design problems. How do you track millions of compounds down to the compound level, if they are made in large batches bound to beads "solid phase synthesis"? In the past, screening a 1000 compounds against a dozen putative enzyme active sites could take many days of computation because both molecules and receptors are flexible and therefore can adopt numerous shapes. But just as computers get faster, the datasets get much larger! Some companies now try to dock 100Ks compounds against 100s receptors.

There is also a growing industry in technologies for discovering "druggable targets" by experimental or computational methods. While genomics gives us some knowledge of the million proteins than can be produced, it does not directly tell us which are potential targets for therapeutic intervention. It's much more complicated than most people realise. Just because an enzyme can be linked to a disease does not mean that stimulating or inhibiting it will lead to a cure (it might actually make things worse) because of biological feedback mechanisms.

Pharmaceutical companies will transition from an historical level of 300-500 targets to ~1000-5000 "druggable targets." Therefore, there is a lot of potential for projects that attempt to put all or some of biochemistry into a computer and run simulations to help understand the dynamics. Companies such as Entelos and Physiome Sciences focus on particular disease processes (e.g., obesity, Asthma) using pharmacokinetic, or other, models. Physiome Sciences has installed several Sun Enterprise™ servers due to reliability, Oracle performance, and scalability. There are several academic projects, e.g. Gepasi/Copasi project at Virginia Bioinformatics Institute.

Chemoinformatic data can be very heterogeneous: registration information, structure, physical analysis results, textual information, materials management (location, quantities, etc.), related biological data, computational results, etc. It is a mixture of text, integer, floating point, categorical, graphical, image, etc. It also tends to be located on legacy systems, which may be in different physical locations, on different hardware, using different software and data formats. While a single energy calculation may run quickly, the trend is to run calculations on large libraries of compounds, say 100K compounds.

Large "in silico" docking experiments (100K compounds vs. 100K proteins) can require teraflop years to compute using current force fields and methods.

IMPLEMENTATIONS OF COMPUTATIONAL TOOLS

Computational tools are artifacts—usually implemented as software, but sometimes as hardware—that enable biologists to solve very specific and precisely defined problems. For example, an algorithm for gene finding or a database of genomic sequences is a computational tool. As a rule, these tools reinforce and strengthen biological research activities, such as recording, managing, analysing, and presenting highly heterogeneous biological data in enormous quantity.

Computational models apply to specific biological phenomena (e.g., organisms, processes) and are used for several purposes. They are used to test insight; to provide a structural framework into which observations and experimental data can be coherently inserted; to make hypotheses more rigorous, quantifiable, and testable; to help identify key or missing elements or important relationships; to help interpret experimental data; to teach or present system behavior; and to predict dynamical behavior of complex systems. Predictive models provide some confidence that certain aspects of a given biological system or phenomenon are understood, when their predictions are validated empirically.

Coming to grips with the complexity of biological phenomena demands an array of intellectual tools to help manage complexity and facilitate understanding in the face of such complexity. In recent years, it has become increasingly clear that many biological phenomena can be understood as performing information processing in varying degrees; thus, a computational perspective that focuses on information abstractions and functional behavior has potentially large benefit for this endeavour.

REFERENCES

James M. Bower (Editor), Hamid Bolouri (Editor) *Computational Modeling of Genetic and Biochemical Networks*. MIT Press. 2001.

Pavel A. Pevzner. *Computational Molecular Biology*. MIT Press. 2000.

Waterman M. S., *Introduction to Computational Biology*. Chapman Hall. 1995.

3

Protein Structure Prediction

Protein structure prediction is one of the most important goals pursued by bioinformatics and theoretical chemistry. Its aim is the prediction of the three-dimensional structure of proteins from their amino acid sequences, sometimes including additional relevant information such as the structures of related proteins. In other words, it deals with the prediction of a protein's tertiary structure from its primary structure. Protein structure prediction is of high importance in medicine (for example, in drug design) and biotechnology (for example, in the design of novel enzymes).

Protein sequence databases are categorised as primary, composite or secondary. The primary structures of a protein are its amino acid sequences; these are stored in primary databases as linear alphabets that denote the constituent residues. The secondary structure of a protein corresponds to regions of local regularity (e.g., alpha helices and beta strands), which, in sequence alignments, are often apparent as well-conserved motifs; these are stored in secondary databases as patterns (e.g., regular expressions, fingerprints, blocks, profiles, etc.). The tertiary structure of a protein arises from the packing of its secondary structure elements, which may form discrete domains within a fold (a, b, c), or may give rise to autonomous folding units or modules; complete folds, domains and modules are stored in structure databases as sets of atomic co-ordinates.

PRIMARY PROTEIN DATABASES

Primary databases contain over 300,000 protein sequences and function as a repository for the raw data. Some more common repositories, such as SWISS-PROT and PIRInternational, annotate the sequences as well as describe the proteins' functions, its domain structure and post-translational modifications. Composite databases such as OWL and the NRDB compile and filter sequence data from different primary databases to produce combined non-redundant sets that are more complete than the individual databases and also include protein sequence data from the translated coding regions in DNA sequence databases.

Secondary databases contain information derived from protein sequences and help the user determine whether a new sequence belongs to a known protein family. One of the most popular is PROSITE, a database of short sequence patterns and profiles that characterise biologically significant sites in proteins. PRINTS expand on this concept and provide a compendium of protein fingerprints—groups of conserved motifs that characterise a protein family. Motifs are usually separated along a protein sequence, but may be contiguous in 3D-space when the protein is folded. By using multiple motifs, fingerprints can encode protein folds and functionalities more flexibly than PROSITE. Finally, Pfam contains a large collection of multiple sequence alignments and profile Hidden Markov Models covering many common protein domains. Pfam-A comprises accurate manually compiled alignments while Pfam-B is an automated clustering of the whole SWISS-PROT database. These different secondary databases have recently been incorporated into a single resource named InterPro.

The Protein Sequence Database was developed at the US National Biomedical Research Foundation (NBRF) in the early 1960s by Margaret Dayhoff as a collection of sequences for investing evolutionary relationships among proteins. Since 1988, the Protein Sequence Database has been maintained collaboratively by PIR-International, an association of macromolecular sequence data collection centers: the consortium includes the Protein Information Resource (PIR) at the NBRF, the International Protein Information Database of Japan (JIPID), and the Martinsried Institute for Protein Sequences (MIPS).

The PIR database is split into 4 distinct sections:

— PIR1: contains fully classified and annotated entries;

— PIR2: includes preliminary entries, which have not been thoroughly reviewed and may contain redundancy;

— PIR3: contains unverified entries and

— PIR4: entries are Conceptual translations of art factual nucleotide sequences.

Martinsried Institute for Protein Sequences (MIPS)

The Martinsried Institute for Protein Sequences collect and processes sequence data for the tripartite PIR- International Protein Sequence Database project. The database is distributed with PATCHx, a supplement of unverified protein sequences from external sources.

SWISS-PROT

SWISS-PROT was produced collaboratively by the Department of Medical Biochemistry at the University of Geneva and the EMBL from its conception in 1986. After 1994, the collaboration moved to EMBL's UK outstation, the EBI. In April 1998 further change saw a move to The Swiss Institute of Bioinformatics (SIB); hence the database is now maintained collaboratively by SIB and EBI/EMBL.

The database endeavors to provide high-level annotations, including descriptions of the function of the protein, and of the structure of its domains, its post-translational modifications, variants, and so on. SWISS-PROT aims to be minimally redundant, and is interlinked to many other resources. In 1998, a computer-annotated supplement of SWISS-PROT was created, termed TrEMBL.

TrEMBL (Translated EMBL) was created in 1996. The database benefits from the SWISS-PROT format, and contains translations of all coding sequences (CDS) in EMBL. TrEMBL has two main sections; designated SP-TrEMBL and REM-EMBL.

1. *SP-TrEMBL*: (SWISS-PROT TrEMBL) contains entries that will eventually be incorporated into SWISS-PROT, but that have not yet been manually annotated.
2. *REM-TrEMBL*: contains sequences that are not destined to be included in SWISS-PROT; these include immunoglobins and T-cell receptors, fragments of fewer than eight amino acids, synthetic sequences, patented sequences and codon translations that do not encode real proteins.

TrEMBL was designed to address the need for a well-structured SWISS-PROT- like resource that would allow very rapid access to sequence data from the genome projects, without having to compromise the quality of SWISS-PROT itself by incorporating sequences with insufficient analysis and annotaton.

Structure of a sequence entry: The entries in the SWISS-PROT database are structured so as to be usable by human readers as well as by computer programs. Wherever possible, symbols familiar to biochemists, protein chemists and molecular biologists are used. Each sequence entry is composed of lines. Different types of lines, each with their own format, are used to record the various data that make up the entry. A sample sequence entry is shown below.

A protein sequence can be retrieved from SWISS-PROT by using the following the URL:

http://us.expasy.org/sprot/

The search can be done by selecting a database of Swiss-prot/TrEMBL for a protein of your interest, e.g., actin or kinase or protease etc. Then you will get a list of entries from the database from which you can choose your sequence by clicking the hyperlink and this will give you the following details regarding the particular protein. The information is divided into different subsections like Entry information, Name and origin of the protein, references, comments, copyright, cross-references, keywords, features and sequence information. For further analysis of the sequence you can get the sequence in FASTA format, just by clicking to the hyperlink on FASTA. This will open a new window with the sequence in FASTA format.

The sequence taken in FASTA format from Swiss-prot can be used for further genome and protein analysis like, pairwise and multiple sequence alignment, BLAST searches where a peptide sequence is an input, motif and domain analysis, secondary structure prediction, protein modeling, protein threading etc.

NRL- 3D

The NRL-3D database is produced by PIR from sequences extracted from the Brookhaven Protein Databank (PDB). The titles and bibliographic sources of the entries conform to the nomenclature standards used in the PIR. Bibliographic references and MEDLINE cross-references are included, together with secondary structure, active site, binding site and modified site annotations, and details of experimental methods, resolution and key words. NRL-3D is a valuable source, as it makes the sequence information in the PDB available both for keyword interrogation and for similarity searches. The database may be searched using the ATLAS retrieval system; a multidatabase information retrieval program specifically designed to access macromolecular sequence databases.

Of the protein sequence databases, NRL-3D is the least comprehensive because it reflects only the contents of the PDB, yet it has the advantage of relating directly to structural information. PIR is the most comprehensive resource, but the quality of its annotations is still relatively poor even in PIR1. Swiss-prot on the other hand, is a highly structured database, that provides excellent annotations, but its sequence coverage is poor compared to PIR. Choosing the right database to search can seem an impossible choice; so is it, perhaps, better to search them all?

COMPOSITE PROTEIN SEQUENCE DATABASES

One solution to the problem of proliferating primary databases is to compile a composite, i.e. a database that amalgamates a variety of different primary sources. Composite database renders sequence searching much more efficient, because they obviate the need to interrogate multiple resources.

Different strategies can be used to create composite resources. The final product depends on the chosen data sources and the criteria used to merge them; for example, a composite resource will be non-identical if it eliminated only identical sequence copies during the amalgamation process; but if both identical and highly similar sequences are ejected (e.g., those entries that differ by only one residue, such as a leading methionine residue), then the resulting database will be more truly non-redundant. The choice of different sources and the application of different redundancy criteria have led to the emergence of different composites, each of which has its own particular format.

1. *NRDB*:- (Non-Redundant Database) is built locally at the NCBI. The database is a composite of Genpept (derived from automatic GenBank CDs translations), PDB sequences, Swiss-prot, Spupdate, PIR and GenPeptudate (the daily updates of Genpept). The

database is thus comprehensive and contains up-to-date information. However, strictly speaking, it is not non-redundant, but non-identical, i.e., only identical sequence copies are removed from the resource. This rather simplistic approach leads to a number of problems:

i. Multiple copies of the same protein are retained in the database as a result of polymorphism and minor sequencing errors.

ii. Incorrect sequences that have been amended in Swiss-prot are reintroduced when retranslated from the DNA.

iii. Numerous sequences are incorporated as full entries of existing fragments. As a result, the contents of NRDB are both error-prone and, in spite of it's name, redundant. Nrdb is the default database of the NCBI BLAST services.

2. *OWL*:—This is a non-redundant protein sequence database built at the University of Leeds in collaboration with the Daresbury Laboratory in Warrington. The database is a composite of four major primary sources: Swiss-prot, PIR1-4, GenBank (CDs translations) and NRL-3D. The sources are assigned a priority with regard to their level of annotation and sequence validation. Swiss-prot has the highest priority; so all others are compared against it during the amalgamation procedure.

This process eliminates both identical copies of sequences and those containing single amino acid differences, leading to a compact (and efficient) resource for sequence comparisons. Nevertheless, the database differs from many of the same problems as NRDB like some sequencing errors and retranslations of incorrect sequences in GenBank are retained. Since OWL is only released on a 6-8 weekly basis, it suffers the further drawback of not being up-to-date. BLAST services for OWL are available from the UK EMBnet National Node, SEQNET, and from the UCL specialist Node.

3. *MIPSX*: This is a merged database produced at the Max-Plank Institute in Martinsried. The database contains information from the following resources- PIR1-4, MIPS preliminary entries, MIPSOWN, MIPS/ PIR preliminary entries, PIRMOD; MIPS preliminary translations, MIPSTrn; MIPS yeast entries, MIPSH; NRL-3D; SWISS-PROT; EMTrans, an automatic translation of EMBL; GBTrans-translated GenBank dentries; Kabat and PseqIP. Sequences that are identical either within or between them are removed, leaving only unique copies. In addition, all subsequences (i.e., sequences completely contained within others) are removed.
4. *SWISS-PROT+ TrEMBL*: At the EBI, the combination of SWISS-PROT and TrEMBL provides a resource that is both comprehensive and 'minimally' redundant. This database had the advantage of containing fewer errors than do those mentioned above, yet it is still not truly non-redundant (in mid-1997, it was estimated that around 30% of the combined total Swiss-prot and TrEMBL was non-unique). To reduce error rates and redundancy levels further will require increasing levels of human intervention and /or the future development of expert database management systems. Swiss-prot and TrEMBL can be reached by means of the SRS sequence retrieval system on the EBI web server.

Ultimately, the choice of database depends on what factors are considered most important fro the job in hand and possibly on whose web server responds most quickly. OWL has the advantage of being fully indexed- means, that the database has been designed for use with a query language, which allows its contents to be rapidly interrogated and manipulated in a variety of different ways. By contrast, NRDB is not available for complex querying, but is useful for up-to-date sequence searches because it contains daily updates of Genpept and weekly updates of Swiss-prot.

SECONDARY DATABASES

There are many secondary (or pattern) databases, so-called because they contain the fruits of analyses of the sequences in the primary sources. Because there are several different primary databases, and a variety of ways of analysing protein sequences, the information housed in each of the secondary resources is different and their formats reflect these disparities. The type of information stored in each of the secondary databases is different. Yet these resources have arisen from a common principle: namely, that homologous sequences may be gathered together in multiple alignments, within which are conserved regions that show little or no variation between the constituent sequences. These conserved regions, or motifs, usually reflect some vital biological role. i.e., they are somehow crucial to the structure or function of the proteins.

PROSITE

PROSITE is a database of protein families and domains. It is based on the observation that, while there is a huge number of different proteins, most of them can be grouped, on the basis of similarities in their sequences, into a limited number of families. Proteins or protein domains belonging to a particular family generally share functional attributes and are derived from a common ancestor. It is apparent, when studying protein sequence families, that some regions have been better conserved than others during evolution. These regions are generally important for the function of a protein and/or for the maintenance of its three- dimensional structure.

By analysing the constant and variable properties of such groups of similar sequences, it is possible to derive a signature for a protein family or domain, which distinguishes its members from all other unrelated proteins. A pertinent analogy is the use of fingerprints by the police for identification purposes. A fingerprint is generally sufficient to identify a given individual. Similarly, a protein signature can be used to assign a newly sequenced protein to a specific family of proteins and thus to formulate hypotheses about its function. Prosite is developed by a largely manual process of seeking the pattern that best fit particular protein families and functions. For instance, two patterns are associated with the serine protease family are given below:

1. [LIVM]-[ST]-A- [STAG]-H-C

2 [DNSTAGC]-[GSTAPIMVQH]-x [LIVMFYWH]- (2)-G- [DE]-S-G- [GS]-[SAPHV]- PA- [LIVMFYSTANQH]

Within these patterns, square brackets indicate sets of possible residues that can occur in a particular position in sequences from the associated family. The first pattern represents a sequence of six residues and should be read as 'one of the residues L, I, V or M followed by S or T, then A then one of S, T, A or G, then H, then C. The feature x (2) means two residues of any type, and not necessarily of the same type. In other patterns the x notation may be extended further, for instance x (2,4) indicates that between two and four residues of any type may be present at a particular position in the family. Another extension of the notation is the use of curly brackets, for instance {LT} means that the position can be occupied by any residue except L or T. PROSITE patterns are most commonly derived from multiple alignments.

Prosite patterns differ from consensus sequences in that they tend to be much shorter than the total sequence length, and that they give a means of describing a set of accepted residues in a multiple alignment column. The patterns can be useful in assigning distant homologs to sequence families, when for instance; all that remains of the sequence similarity is limited to a few important residues around the key functional machinery of the protein. They can therefore also be indicative of shared biological function. As well as patterns associated with family membership, PROSITE contains generic patterns associated with processes like post-translational modification (glycosylation, phosphorylation, etc.).

Limitations of PROSITE:

1. There is a possibility of false positives.
2. While they allow the description of variation at a particular position, they have no way of attaching probabilities to the variation.

For instance, [LIVM] says that a position might be L, I, V or M, but it does not say that perhaps L occurs in 90% of sequences in the family and I, V or M only in the other 10%. In order to overcome some of the difficulties associated with PROSITE patterns, the PROSITE database now contains sequence profiles. These attempts to describe linger sequence segments (usually complete domains) than the patterns. There are also several other databases that are complementary tools to PROSITE in sequence analysis, including PRINTS and BLOCKS, Pfam and SMART.

PROSITE currently contains patterns and profiles specific for more than a thousand protein families or domains. Each of these signatures comes with documentation providing background information on the structure and function of these proteins.

PRINTS AND BLOCKS

They are closely related, each representing protein families in terns of multiply aligned ungapped segments derived from the most highly conserved regions in a group of proteins or protein family. Such multiply aligned ungapped segments are termed blocks (in BLOCKS) or motifs (in PRINTS). In PRINTS, a set of such motifs represented a family is called a fingerprint. A motif represents any conserved element of a sequence alignment: it is a local alignment corresponding to a region whose function or structure is known, or its significance may be unknown. It is sufficient that it is conserved, and is hence likely to be predictive of any subsequent occurrence of such a structural/functional region in any other protein sequence.

A fingerprint is a set of motifs used to predict the occurrence of similar motifs, either in an individual sequence or in a database. Fingerprints are refined, i.e. their diagnostic potency is enhanced, by iterative scanning of the OWL composite sequence database. Database searches with such aligned motifs are essentially frequency scans: i.e., in their most basic application, no secondary structure information, similarity data, or weighting scheme of any description is used to improve discrimination power.

The fingerprinting method relies on the fact that, in any protein family, only parts of a sequence are held in common—these usually relate to key functional regions or to core structural elements of the fold. The starting point for fingerprint definition is thus multiple sequence alignment, which we achieve using the SOMAP, XALIGN or VISTAS manual alignment programs. Only small numbers need to be included in the initial alignment because the method itself is designed to add to the alignment with each database scan. Once a motif, or set of motifs, has been identified, the conserved regions are excised in the form of local alignments. There are no rules regarding the juxtaposition of such motifs, other than that they should not substantially overlap: thus motifs may occur immediately adjacent to one another, or they may be separated by any distance along the length of the alignment.

Independent database scans are made with each aligned motif, resulting in the production of a set of hit lists, one for each motif. The hit lists are then analysed, or correlated, to determine which sequences in the database have matched with all elements of the fingerprint, and which have only matched with part of it. Only those sequences that match with all elements are regarded as true matches. If the search has worked well, the true set will contain more sequences than did the original alignment.

The additional sequence data from the new true set is then used to generate another set of aligned motifs, and the database is searched again—where the family being diagnosed is very large, redundant motifs are removed from the alignments prior to the next scan. This process is repeated until convergence, the point at which the true set remains constant between successive scans.

The final aligned motifs from this iterative procedure constitute the refined fingerprint that is entered into the PRINTS database. Good fingerprints find all true matches that exist in OWL; they exhibit clear discrimination cut-offs, and include little or no noise. Occasionally, cut-offs may be difficult to assess: this is usually the result of subfamily identification, where a sequence subset has been recognised that is characterised by only part of a fingerprint; but it may also result from the use of only 2 or 3 motifs in the fingerprint—discrimination power improves with the number of motifs used.

The PRINTS database is of very high quality, having been created with a great deal of manual effort, and contains extensive annotation and description for the protein family and function concerned. PRINTS is a compendium of protein fingerprints. A fingerprint is a group of conserved motifs used to characterise a protein family; its diagnostic power is refined by iterative scanning of a SWISS-PROT/TrEMBL composite. Usually the motifs do not overlap, but are separated along a sequence, though they may be contiguous in 3D-space. Fingerprints can encode protein folds and functionalities more flexibly and powerfully than can single motifs, full diagnostic potency deriving from the mutual context provided by motif neighbors.

The original version of BLOCKS was created by automatic means, but now many databases (including PRINTS) are available in BLOCKS format. The motifs within these databases typically cover larger regions of the sequence than PROSITE patterns, and unlike PROSITE, matching of motifs in sequences usually takes account of amino acid substitution matrices, this not requiring exact matches to a fixed patterns. Search engines for the databases are available from:

http://bioinfo.man.ac.uk/dbbrowser/PRINTS

http:// www.blocks.fhcrc.org/

Search

To search, simply enter a query-type the words you want to search for in the 'search term box' (or follow the instructions from your client as to how to enter a search). If you enter two or more words, than any document that contains ANY of the words will be found. (For example, if you search for ice cream, then any document with either ice or cream will be on the list.) If you want only documents containing BOTH ice AND cream, then use AND between the words; your query would be: ice and cream. The list below for a full description of searching options and their effect. The result of this search will be a list of the documents, which satisfy your search request (or a reply that indicates that NO documents match). The list is ordered by 'relevancy rank' (highest first), which very roughly corresponds to the number of times the word you searched for occurred in the document versus the size of the document.

The following is a list of all the search options available:

Right-hand truncation (stemming) queries-Expand Your Search

The query 'astro' will find documents containing the words 'astronomy' as well as 'astrophysics'. If you are not sure of the spelling or form of the word used in the documents, use this.

'And' queries-Narrow Your Search

The query 'red and blue' will find the intersection of all the documents containing the words 'red', and 'blue'. The use of 'and' limits your retrieval.

'Or' queries-Expand Your Search

The query 'red or blue' will find the union of all the documents containing the words 'red' and 'blue'. The use of 'or' increases your retrieval.

'Not' queries-Narrow Your Search

The query 'red not green' will find all the documents containing the word 'red', and excluding the documents containing the word 'green'. The use of 'not' limits your retrieval.

Nested queries-Complicated Searches

The query '(red and green) or blue not pink' will find the union of all the documents containing the words 'red', and 'green'. It will then add (union) all documents containing the word 'blue'. Finally, it will exclude all documents containing the word 'pink'. Release 35.0 of PRINTS contains 1750 entries, encoding 10,626 individual motifs. Overall, the database is still relatively small, largely because the detailed annotation of fingerprints is so time-consuming.

Nevertheless, the additional, unique information contained in PRINTS makes this database a useful adjunct to PROSITE, Pfam, Blocks, etc. PRINTS is released in major and minor versions: minor releases reflect updates, bringing the contents in line with the current version of the source database(s) (formerly, PRINTS was derived from OWL, but since version 23.1, the database has changed to SWISS-PROT and TrEMBL sources); major releases denote the addition of new material to the resource.

Pfam

Pfam is a database of multiple alignments of protein domains or conserved protein regions. The alignments represent some evolutionary conserved structure, which has implications for the protein's function. Profile hidden Markov models (profile HMMs) built from the Pfam alignments can be very useful for automatically recognising that a new protein belongs to an existing protein family, even if the homology is weak. Unlike standard pairwise alignment methods (e.g. BLAST, FASTA), Pfam HMMs deal sensibly with multidomain proteins.

Pfam is formed in two separate ways. Pfam-A are accurate human crafted multiple alignments, whereas Pfam-B is an automatic clustering of the rest of a nonredundant protein database derived from the PRODOM database.

Protein Domain Families

Many proteins are constructed from more than one domain and some domains are common to many protein families. For instance, (SH2 and SH3) domains appear in many proteins associated with signalling, and pleckstrin homology (PH) domains appear in many proteins that bind phospholipids. Genetic events can result in domain swapping, domain duplication, and loss and gain of domains. Acquiring new domains with specific functions allows proteins to acquire new and more complex functions very quickly. The Prodom is a database of protein domain family sequences.

It is created automatically from the database of known protein sequences, using pair-wise sequence comparison with the BLAST tools followed clustering together of related sequence segments from different proteins into domain families. It attempts to identify independent sequence segments with similarity detectable at the sequence level, which are shared by many proteins. These are identified with domains. The server allows a protein sequence to be compared to the database of domain sequences, in order to identify shared domains.

Sequence Profile

Sequence profile is a way of describing related sequences from a protein domain family. They have been adopted by the PROSITE database to supplement the PROSITE pattern type entries. Their principal advantage is that they describe a complete domain sequence, including the likelihood of observing each amino acid, along with the likelihood of insertion and deletion, at each position in the sequence. Each multiple alignment column in the aligned family sequences is associated with a set of 20 scores, one for the appearance of each amino acid at that position. Observed amino acids tend to score highly, but the profile is derived in such a way that amino acids that substitute for the observed amino acids score reasonably highly as well. A new sequence can be aligned to a profile so that its amino acid residues match with columns where they obtain high scores. A high total alignment score indicates that the sequence is likely to belong to the family.

The Pfam database is a collection of protein domain family multiple alignments and HMMs maintained at the Sanger Centre. Pfam has elucidated the full domain structure of the protein, including two PH domains and a single DEP (Disheveled, Eg1-10, Pleckstrin) domain. Another useful resource that will carry out this type of analysis is SMART (Simple Modular Architecture Research Tool).

A recent and very valuable development is the integration of several of the protein family resources into a resource called InterPro. This unites information from many of the resources, including PROSITE, PRINTS, Pfam, Prodom and SMART, allowing integrated search and sequence analysis.

InterPro: http://www.ebi.ac.uk/interpro/

STRUCTURAL DATABASES

The Protein Data Bank, PDB, provides a primary archive of all 3D structures for macromolecules such as proteins, RNA, DNA and various complexes. As the information provided in individual PDB entries can be difficult to extract, PDB sum provides a separate Web page for every structure in the PDB displaying detailed structural analyses, schematic diagrams and data on interactions between different molecules in a given entry. Three major databases classify proteins by structure in order to identify structural and evolutionary relationships:

CATH, SCOP, and FSSP databases. All comprise hierarchical structural taxonomy where groups of proteins increase in similarity at lower levels of the classification tree. In addition, numerous databases focus on particular types of macromolecules. These include the Nucleic Acids Database, NDB, for structures related to nucleic acids, the HIV protease database for HIV-1, HIV-2 and SIV protease structures and their complexes, and ReLiBase for receptor-ligand complexes.

PDB

The PDB is a database consisting of a set of ASCII files, each containing the Cartesian atomic coordinates describing the three-dimensional structure of a protein, nucleic acid or other bio-macromolecule. Most were determined by X-ray crystallography, and some by NMR. These files are linked by a relational database protocol with a Web-based search front-end.

Historically, the PDB was founded and developed at the Brookhaven National Lab, in the center of Long Island, NY USA, for over 25 years. In 1999 it had to move to the RCSB at Rutgers University, NJ USA. As part of this move the Web interface took on a new look. However, non-American mirror sites, such as the Weizmann mirror site have preserved the old Brookhaven interface and browsers that most users are familiar with. The principal method to access and search the holdings of the PDB is via its Web interface, using your favorite Internet browser. Direct your browser either to the main RSCB pdb location at URL:

http://www.rcsb.org

Or the local mirror copy, which uses the old Brookhaven interface and browser, at URL:

http://pdb.weizmann.ac.il

In either place, the next step is to select 'Search the PDB' to bring up the browser. The Brookhaven interface supports two search tools, the full PDB Browser, and the simplified PDB-lite browser for casual users and lab biologists. Whichever browser you use, you fill in the blanks with appropriate search questions, based on the header lines in the PDB files. Options exist to search by PDB id-code (the unique 4-character id the PDB assigns to each entry), or by sequence in a FASTA-like interface (which of-course returns only sequences with structures in the PDB).

If only a single structure is found, the browser returns an Atlas Page for the structure, including a summary of its header information, and options to download some or all the entry, to display the protein in various ways (Rasmol, Chime, VRML, GIF/JPEG pictures), and to locate related information on other databases. (While the Brookhaven atlas has everything on one page, the RSCB atlas is fragmented into a few pages). If the search returns several entries, these are returned in a combined atlas page, with an abbreviated entry for each structure found. You can then choose which structure(s) to examine in more detail, or modify your search question to make it more specific.

Besides searching the PDB, the Web sites contain a wealth of information about the PDB, including policy statements, a format guide, and a quarterly newsletter that you can subscribe to. There are also some links to related databases and some free software you can download. The number of structures deposited in the PDB in the year 2002 is 1764 and total as on date is 18606 Proportion of "new folds" and "old folds" for a given year as a number of protein chains based on the CE algorithm.

Note: A chain fold was considered old if it was similar to one of selected chains representing new folds according to the following criteria: z-score >= 4.0, RMSD < 3.0Å, number of aligned positions >= 70% of the length of this chain. Otherwise the chain fold was considered new.

STRUCTURAL CLASSIFICATION OF PROTEINS (SCOP)

The SCOP database aims to provide a detailed and comprehensive description of the structural and evolutionary relationships between all proteins whose structure is known, including all entries in the Protein Data Bank (PDB). It is available as a set of tightly linked hypertext documents which make the large database comprehensible and accessible. In addition, the hypertext pages offer panoply of representations of proteins, including links to PDB entries, sequences, references, images and interactive display systems.

World Wide Web URL http://scop.mrc-lmb.cam.ac.uk/scop/ is the entry point to the database (MRC site). Existing automatic sequence and structure comparison tools cannot identify all structural and evolutionary relationships between proteins. The SCOP classification of proteins has been constructed manually by visual inspection and comparison of structures, but with the assistance of tools to make the task manageable and help provide generality. The job is made more challenging-and theoretically daunting-by the fact that the entities being organised are not homogeneous: sometimes it makes more sense to organise by individual domains, and other times by whole multi-domain proteins.

Categorisation

Proteins are classified to reflect both structural and evolutionary relatedness. Many levels exist in the hierarchy, but the principal levels are family, super family and fold, described below. The exact position of boundaries between these levels is to some degree subjective. The evolutionary classification is generally conservative: where any doubt about relatedness exists, we made new divisions at the family and super family levels. Thus, some researchers may prefer to focus on the higher levels of the classification tree, where proteins with structural similarity are clustered.

The different major levels in the hierarchy are:

Family

Proteins clustered together into families are clearly evolutionarily related. Generally, this means that pairwise residue identities between the proteins are 30% and greater. However, in some cases similar functions and structures provide definitive evidence of common descent in the absence of high sequence identity; for example, many globins form a family though some members have sequence identities of only 15%.

Super family

Proteins that have low sequence identities, but whose structural and functional features suggest that a common evolutionary origin is probable are placed together in super families. For example, actin, the ATPase domain of the heat shock protein, and hexakinase together form a superfamily.

Fold

Proteins are defined as having a common fold if they have the same major secondary structures in the same arrangement and with the same topological connections. Different proteins with the same fold often have peripheral elements of secondary structure and turn regions that differ in size and conformation. In some cases, these differing peripheral regions may comprise half the structure. Proteins placed together in the same fold category may not have a common evolutionary origin: the structural similarities could arise just from the physics and chemistry of proteins favoring certain packing arrangements and chain topologies.

Usage

We hope that SCOP will have broad utility that will attract a wide range of users. Experimental structural biologists may wish to explore the region of "structure space" near their proteins of current research, while theoreticians will likely find it most useful to browse the wide range of protein folds currently known. Molecular biologists may find the classification helpful because the categorisation assists in locating proteins of interest and the links make exploration easy. We also hope that SCOP will find pedagogical use, for it organises structures in an easily comprehensible manner and makes them accessible from even a simple personal computer.

CATH DATABASE

The CATH database is a hierarchical domain classification of protein structures in the Brookhaven protein databank. All non-protein, model, and "C-alpha only" structures are not classified in CATH. Only crystal structures solved to resolution better than 3.0 angstroms are considered, together with NMR structures. This filtering of the Brookhaven databank is performed using the program SIFT. There are four major levels in this hierarchy; Class, Architecture, Topology (fold family) and Homologous super family. Each level is described below, together with the methods used for assigning structures to a specific family.

Domains

Multidomain proteins are subdivided into their constituent domains using a consensus procedure, based on three independent algorithms for domain recognition, PUU and DOMAK. This currently allows approximately 53% of the proteins (i.e. those for which these algorithms agree) to be defined as single or multidomain proteins automatically. The remaining structures are assigned domain definitions manually, by choosing what was determined to be the best assignment made by one of the algorithms, a new assignment, or an alternative assignment obtained from the literature. The multidomain proteins are then split into their separate domains. All the classification is performed on individual protein domains.

CATH hierarchy

Class, C-level

Class is determined according to the secondary structure composition and packing within the structure. It can be assigned automatically for over 90% of the known structures using the method of Michie et al. For the remainder, manual inspection is used and where necessary information from the literature taken into account. Three major classes are recognised; mainly alpha, mainly-beta and alpha-beta. This last class (alpha-beta) includes both alternating alpha/beta structures and alpha+beta structures, as originally defined by Levitt and Chothia. A fourth class is also identified which contains protein domains, which have low secondary structure content.

Architecture, A-level

This describes the overall shape of the domain structure as determined by the orientations of the secondary structures but ignores the connectivity between the secondary structures. It is currently assigned manually using a simple description of the secondary structure arrangement e.g. barrel or 3-layer sandwich. Reference is made to the literature for wellknown architectures (e.g the beta-propellor or alpha four helix bundle). Procedures are being developed for automating this step.

Topology (Fold family), T-level

Structures are grouped into fold families at this level depending on both the overall shape and connectivity of the secondary structures. This is done using the structure comparison algorithm SSAP. Parameters for clustering domains into the same fold family have been determined by empirical trials throughout the databank, Structures, which have a SSAP score of 70 and where at least 60% of the larger protein matches the smaller protein, are assigned to the same T level or fold family.

Some fold families are very highly populated particularly within the mainly-beta 2-layer sandwich architectures and the alpha-beta 3-layer sandwich architectures. In order to appreciate the structural relationships within these families more easily, they are currently subdivided using a higher cutoff on the SSAP score (75 for some mainly-beta and alphabeta families, 80 for some mainly-alpha families, together with a higher overlap requirement (70%)).

Homologous Superfamily, H-level

This level groups together protein domains, which are thought to share a common ancestor and can therefore be described as homologous. Similarities are identified first by sequence comparisons and subsequently by structure comparison using SSAP. Structures are clustered into the same homologous super family if they satisfy one of the following criteria:

— Sequence identity >= 35%, 60% of larger structure equivalent to smaller
— SSAP score >= 80.0 and sequence identity >= 20%
— 60% of larger structure equivalent to smaller
— SSAP score >= 80.0, 60% of larger structure equivalent to smaller, and domains which have related functions.

Sequence families, S-level

Structures within each H-level are further clustered on sequence identity. Domains clustered in the same sequence families have sequence identities >35% (with at least 60% of the larger domain equivalent to the smaller), indicating highly similar structures and functions.

GENE EXPRESSION DATABASES

The most recent sources of genomic-scale data have been from expression experiments, which quantify the expression levels of individual genes. These experiments measure the amount of mRNA or protein products that are produced by the cell. For the former, there are three main technologies: the cDNA microarray, Affymatrix GeneChip and SAGE methods. The first method measures relative levels of mRNA abundance between different samples, while the last two measure absolute levels.

Most of the effort in gene expression analysis has concentrated on the yeast and human genomes and as yet, there is no central repository for this data. For yeast, the Young, Church and Samson datasets use the GeneChip method, while the Stanford cell cycle, diauxic shift and deletion mutant datasets use the microarray. Most measure mRNA levels throughout the whole yeast cell cycle, although some focus on a particular stage in the cycle. For humans, the main application has been to understand expression in tumour and cancer cells. The Molecular Portraits of Breast Tumours, Lymphoma and Leukaemia Molecular Profiling projects provide data from microarray experiments on human cancer cells. The technologies for measuring protein abundance are currently limited to 2D gel electrophoresis followed by mass spectrometry. As gels can only routinely resolve about 1,000 proteins, only the most abundant can be visualised. At present, data from these experiments are only available from the literature.

2D-PAGE Database

SWISS-2DPAGE is an annotated two-dimensional polyacrylamide gel electrophoresis (2-D PAGE) and SDS-PAGE database established in 1993 and maintained collaboratively by the Central Clinical Chemistry Laboratory of the Geneva University Hopital and the Swiss Institute of Bioinformatics (SIB).

The SWISS-2DPAGE database assembles data on proteins identified on various 2-D PAGE and SDS-PAGE maps. Each SWISS-2DPAGE entry contains textual data on one protein, including mapping procedures, physiological and pathological information, experimental data (isoelectric point, molecular weight, amino acid composition, peptide masses) and bibliographical references. In addition to this textual data, SWISS-2DPAGE provides several 2-D PAGE and SDS-PAGE images showing the experimentally determined location of the protein, as well as a theoretical region computed from the sequence protein, indicating where the protein might be found in the gel. Cross-references are provided to Medline and other federated 2-DE databases (COMPLUYEAST-2DPAGE, ECO2DBASE, HSC-2DPAGE, PHCI-2DPAGE, PMMA-2DPAGE, Siena-2DPAGE, YEPD) and to SWISS-PROT, which provides many links to other molecular databases (EMBL, Genbank, PROSITE, OMIM, etc).

The entries in the SWISS-2DPAGE are text files structured to be readable by human as well as by computer programs. The explanations, descriptions, classifications and comments are in everyday English. However, symbols familiar to biochemists, chemists and molecular biologists are also used. Each entry corresponds to one protein and is composed of lines. Different types of lines, each with their own format, are used to record the various data which make up the entry.

PROTEIN STRUCTURE MODELLING

The Protein Structure Modeling set includes a whole-proteome, high-throughput protein structure modeling system to predict protein structure from the amino acids sequence of the protein and structures of known proteins. The set is also capable of comprising several algorithms for identifying protein signatures, automated clustering of proteins based on structure and identifying unique protein signature based on structure. In addition, the set can help identify biologically important residues in protein.

The Protein Structure Modeling technology has several attributes:

1) Finds similarities between two structures and fragments of protein
2) Allows identification and analysis of structural similarities of proteins that do not have significant amino acid sequence similarity
3) Allows clustering of similar fragments of structures. Such clusters can be used to identify sequence patterns which will represent local structural motifs in proteins

The use of Protein Structure Modeling technology will significantly improve the process of fold recognition and more distant homologs detection. Also in protein structure prediction methods, it will enhance the quality and accuracy of the final 3D protein models especially when small fragments of proteins need to be modeled (loops, deletions, insertions, signature regions).

Comparative Modelling

All current comparative modelling methods consist of five sequential steps. Each steps are used to:

— search for proteins with known 3D structures that are related to the target sequence.

— pick those structures that will be used as templates.

— align their sequences with the target sequence.

— build the model for the target sequence given its alignment with the template structures.

— evaluate the model using a variety of criteria.

If necessary, template selection, alignment and model building can be repeated until a satisfactory model is obtained. Each of the above steps are briefly described below.

Searching for Structures Related to the Target Sequence

Comparative modelling usually starts by searching the PDB of known protein structures using the target sequence as the query. This search is generally done by comparing the target sequence with the sequence of each of the structures in the database. There are three main classes of protein comparison methods that are useful in fold identification.

The first class compares the target sequence with each of the database sequences independently, using pairwise sequence-sequence comparison. The performance of these methods in sequence searching and fold assignments has been evaluated exhaustively. The most popular programs in the class include FASTA and BLAST.

The second class of methods relies on multiple sequence comparisons to improve greatly the sensitivity of the search. The most well known program in this class is PSI-BLAST. Another similar approach that appears to perform even slightly better than PSI-BLAST has been implemented in the program PDB-BLAST. PDB-BLAST begins by finding all sequences in a sequence database that are clearly related to the target and easily aligned with it. The multiple alignment of these sequences is the target sequence profile. Similar profiles are also constructed for all potential template structures. The templates are then found by comparing the target sequence profile with each of the template sequence profiles, using a local dynamic programming method that relies on the common BLOSUM62 residue substitution matrix. These more sensitive fold identification techniques based on sequence profiles are especially useful for finding significant structural relationships when sequence identity between the target and the template drops below 25%.

The third class of methods relies on pairwise comparison of a protein sequence and a protein structure; that is, structural information is used for one of the two proteins that are being compared, and the target sequence is matched against a library of 3D profiles or threaded through a library of 3D folds. These methods are also called fold assignment, threading or 3D template matching.

These methods are especially useful when sequence profiles are not possible to construct because there are not enough known sequences that are clearly related to the target or potential templates. The usefulness of a template should be assessed by evaluation of the corresponding 3D model based on a given template, using methods described below. This approach is optimal because the evaluation of a 3D model is generally more sensitive and robust than the evaluation of an alignment. A good starting point for template searches are the many database search servers on the web.

Template Selection

Once a list of potential templates is obtained using searching methods, it is necessary to select one or more templates that are appropriate for the particular modelling problem. Several factors need to be taken into account when selecting a template. The quality of a template increases with its overall sequence similarity to the target and decreases with the number and length of gaps in the alignment. The simplest template selection rule is to select the structure with the highest sequence similarity to the modeled sequence.

The family of proteins that includes the target and the templates can frequently be organised into sub-families. The construction of a multiple alignment and a phylogenetic tree can help in selecting the template from the subfamily that is closest to the target sequence. The similarity between the "environment" of the template and the environment in which the target needs to be modeled should also be considered. The term "environment" is used here in a broad sense, including everything that is not the protein itself (eg, solvent, pH, ligands, quaternary interactions). If possible, a template bound to the same or similar ligands as the modeled sequence should generally be used.

The quality of the experimentally determined structure is another important factor in template selection. Resolution and R-factor of a crystallographic structure and the number of restraints per residue for an NMR structure are indicative of the accuracy of the structure. For instance, if two templates have comparable sequence similarity to the target, the one determined at the highest resolution should generally be used. The criteria for selecting templates also depend on the purpose of a comparative model. For example, if a protein-ligand model is to be constructed, the choice of the template that contains a similar ligand is probably more important than the resolution of the template.

On the other hand, if the model is to be used to analyse the geometry of the active site of an enzyme, it may be preferable to use a high-resolution template structure. It is not necessary to select only one template. In fact, the use of several templates generally increases the model accuracy. One strength of comparative modelling program MODELLER is that it can combine information from multiple template structures, in two ways.

First, multiple template structures may be aligned with different domains of the target, with little overlap between them, in which case the modelling procedure can construct a homology-based model of the whole target sequence.

Second, the template structures may be aligned with the same part of the target, in which case the modelling procedure is likely to automatically build the model on the locally best template.

In general, it is frequently beneficial to include in the modelling process all the templates that differ substantially from each other, if they share approximately the same overall similarity to the target sequence. An elaborate way to select suitable templates is to generate and evaluate models for each candidate template structure and/or their combinations. The optimised all-atom models are evaluated by an energy or scoring function, such as the Z-score of PROSA. The PROSA Z-score of a model is a measure of compatibility between its sequence and structure. Ideally, the Z-score of the model should be comparable to the Z-score of the template. PROSA Z-score is frequently sufficiently accurate to allow picking one of the most accurate of the generated models. This trial-and-error approach can be viewed as limited threading (ie, the target sequence is threaded through similar template structures).

Alignment of the Target and Template Sequences

To build a model, all comparative modelling programs depend on a list of assumed structural equivalences between the target and template residues. This list is defined by the alignment of the target and template sequences. Although many template search methods will produce such an alignment, it is usually not the optimal target-template alignment in the more difficult alignment cases (eg, at less than 30% sequence identity). Search methods tend to be tuned for detection of remote relationships, not for optimal alignment. Therefore, once the templates are selected, an alignment method should be used to align them with the target sequence.

The alignment is relatively simple to obtain when the target-template sequence identity is above 40%. In most such cases, an accurate alignment can be calculated automatically using standard sequence-sequence alignment methods. If the target-template sequence identity is lower than 40%, the alignment generally has gaps and needs manual intervention to minimise the number of misaligned residues. In these low sequence identity cases, the alignment accuracy is the most important factor affecting the quality of the resulting model. Alignments can be improved by including structural information from the template. For example, gaps should be avoided in secondary structure elements, in buried regions, or between two residues that are far in space.

Some alignment methods take such criteria into account. It is important to inspect and edit the alignment in view of the template structure, especially if the target-template sequence identity is low. A misalignment by only one residue position will result in an error of approximately 4Å in the model because the current modelling methods generally cannot recover from errors in the alignment. When multiple templates are selected, a good strategy is to superpose them with each other first, to obtain a multiple structure-based alignment. In the next step, the target sequence is aligned with this multiple structure-based alignment.

Another improvement is to calculate the target and template sequence profiles, by aligning them with all sequences from a nonredundant sequence database that are sufficiently similar to the target and template sequences, respectively, so that they can be aligned without significant errors. The final target-template alignment is then obtained by aligning the two profiles, not the template and target sequences alone. The use of multiple structures and multiple sequences benefits from the evolutionary and structural information about the templates as well as evolutionary information about the target sequence, and often produces a better alignment for modelling than the pairwise sequence alignment methods.

Model Construction

Assembly of rigid bodies modelling

The first and still widely used approach in comparative

modelling is to assemble a model from a small number of rigid bodies obtained from the aligned protein structures. The approach is based on the natural dissection of the protein structure into conserved core regions, variable loops that connect them, and sidechains that decorate the backbone. For example, the following semi-automated procedure is implemented in the computer program COMPOSER.

— The template structures are selected and superposed.
— The "framework" is calculated by averaging the coordinates of the C_a atoms of structurally conserved regions in the template structures.
— The mainchain atoms of each core region in the target model are obtained by superposing on the framework the core segment from the template whose sequence is closest to the target.
— The loops are generated by scanning a database of all known protein structures to identify the structurally variable regions that fit the anchor core regions and have a compatible sequence.
— The sidechains are modeled based on their intrinsic conformational preferences and on the conformation of the equivalent sidechains in the template structures.
— The stereochemistry of the model is improved either by a restrained energy minimisation or a molecular dynamics refinement.

The accuracy of a model can be somewhat increased when more than one template structure is used to construct the framework and when the templates are averaged into the framework using weights corresponding to their sequence similarities to the target sequence. For example, differences between the model and X-ray structures may be slightly smaller than the differences between the X-ray structures of the modeled protein and the homologs used to build the model. Possible future improvements of modelling by rigid body assembly include incorporation of rigid body shifts, such as the relative shifts in the packing of α-helices and β-sheets.

Coordinate reconstruction modelling

The basis of modelling by coordinate reconstruction is the finding that most hexapeptide segments of protein structure can be clustered into only 100 structurally different classes. Thus, comparative models can be constructed by using a subset of atomic positions from template structures as "guiding" positions, and by identifying and assembling short, all-atom segments that fit these guiding positions. The guiding positions usually correspond to the C_a atoms of the segments that are conserved in the alignment between the template structure and the target sequence.

The all-atom segments that fit the guiding positions can be obtained either by scanning all the known protein structures, including those that are not related to the sequence being modeled, or by a conformational search restrained by an energy function. For example, a general method for modelling by segment matching is guided by the positions of some atoms (usually C_a atoms) to find the matching segments in the representative database of all known protein structures. This method can construct both mainchain and sidechain atoms, and can also model gaps. It is implemented in the program SegMod. Even some sidechain modelling methods and the class of loop construction methods based on finding suitable fragments in the database of known structures can be seen as segment matching or coordinate reconstruction methods.

Satisfaction of spatial restraints modelling

The methods in this class begin by generating many constraints or restraints on the structure of the target sequence, using its alignment to related protein structures as a guide. The procedure is conceptually similar to that used in determination of protein structures from NMR-derived restraints. The restraints are generally obtained by assuming that the corresponding distances between aligned residues in the template and the target structures are similar. These homologyderived restraints are usually supplemented by stereochemical restraints on bond lengths, bond angles, dihedral angles, and non-bonded atom-atom contacts that are obtained from a molecular mechanics force field.

The model is then derived by minimising the violations of all the restraints. This can be achieved either by distance geometry or real-space optimisation. For example, an elegant distance geometry approach constructs all-atom models from lower and upper bounds on distances and dihedral angles. Lower and upper bounds on $C_{a-}C_{a}$ and mainchain-sidechain distances, hydrogen bonds, and conserved dihedral angles were derived for E. coli flavodoxin from four other flavodoxins; bounds were calculated for all distances and dihedral angles that had equivalent atoms in the template structures. The allowed range of values of a distance or a dihedral angle depended on the degree of structural variability at the corresponding position in the template structures.

Distance geometry was used to obtain an ensemble of approximate 3D models, which were then exhaustively refined by restrained molecular dynamics with simulated annealing in water. Some researchers approaches are briefly described below. The approach was developed to use as many different types of data about the target sequence as possible. It is implemented in the computer program MODELLER. The comparative modelling procedure begins with an alignment of the target sequence with related known 3D structures. The output, obtained without any user intervention, is a 3D model for the target sequence containing all mainchain and sidechain nonhydrogen atoms.

In the first step of model building, distance and dihedral angle restraints on the target sequence are derived from its alignment with template 3D structures. The form of these restraints was obtained from a statistical analysis of the relationships between similar protein structures. The analysis relied on a database of 105 family alignments that included 416 proteins of known 3D structure. By scanning the database of alignments, tables quantifying various correlations were obtained, such as the correlations between two equivalent C_a-C_a distances, or between equivalent mainchain dihedral angles from two related proteins. These relationships are expressed as conditional probability density functions (pdf's) and can be used directly as spatial restraints.

Loop modelling

In comparative modelling, target sequences often have inserted residues relative to the template structures or have regions that are structurally different from the corresponding regions in the templates. Thus, no structural information about these inserted segments can be extracted from the template structures. These regions frequently correspond to surface loops. Loops often play an important role in defining the functional specificity of a given protein framework, forming the active and binding sites.

The accuracy of loop modelling is a major factor determining the usefulness of comparative models in applications such as ligand docking. Loop modelling can be seen as a mini protein folding problem because the correct conformation of a given segment of a polypeptide chain has to be calculated mainly from the sequence of the segment itself. However, loops are generally too short to provide sufficient information about their local fold. Even identical decapeptides in different proteins do not always have the same conformation. Some additional restraints are provided by the core anchor regions that span the loop and by the structure of the rest of a protein that cradles the loop.

Although many loop modelling methods have been described, it is still not possible to model correctly and with high confidence loops longer than approximately 8 residues. There are two main classes of loop modelling methods:

(i) the database search approaches, where a segment that fits on the anchor core regions is found in a database of all known protein structures;

(ii) the conformational search approaches. There are also methods that combine these two approaches.

The database search approach to loop modelling is accurate and efficient when a database of specific loops is created to address the modelling of the same class of loops, such as ß-hairpins, or loops on a specific fold, such as the hypervariable regions in the immunoglobulin fold. For example, an analysis of the hypervariable immunoglobulin regions resulted in a series of rules that allowed a very high accuracy of loop prediction in other members of the family. These rules were based on the small number of conformations for each loop, and the dependence of the loop conformation on its length and certain key residues.

There are attempts to classify loop conformations into more general categories, thus extending the applicability of the database search approach for to more cases. However, the database methods are limited by the fact that the number of possible conformations increases exponentially with the length of a loop. As a result, only loops up to 4-7 residues long have most of their conceivable conformations present in the database of known protein structures. Even according to the more optimistic estimate, approximately 30% and 60% of all the possible 8 and 9 residue loop conformations, respectively, are missing from the database. This is made even worse by the requirement for an overlap of at least one residue between the database fragment and the anchor core regions, which means that the modelling of a 5 residue insertion requires at least a 7 residue fragment from the database.

Despite the rapid growth of the database of known structures, there is no possibility to cover most of the conformations of a 9-residue segment in the foreseeable future. On the other hand, most of the insertions in a family of homologous proteins are shorter than 10-12 residues. To overcome the limitations of the database search methods, conformational search methods were developed. There are many such methods, exploiting different protein representations, objective function terms, and optimisation or enumeration algorithms. The search algorithms include the minimum perturbation method, molecular dynamics simulations, genetic algorithms, Monte Carlo and simulated annealing, multiple copy simultaneous search, self-consistent field optimisation, and an enumeration based on the graph theory.

The loop modelling module in MODELLER implements the optimisation-based approach. The main reasons are the generality and conceptual simplicity of energy minimisation, as well as the limitations on the database approach imposed by a relatively small number of known protein structures. Loop prediction by optimisation is applicable to simultaneous modelling of several loops and loops interacting with ligands, which is not straightforward for the database search approaches. Loop optimisation in MODELLER relies on conjugate gradients and molecular dynamics with simulated annealing.

The pseudo energy function is a sum of many terms, including some terms from the CHARMM-22 molecular mechanics force field and spatial restraints based on distributions of distances and dihedral angles in known protein structures. The method was tested on a large number of loops of known structure, both in the native and nearnative environments. In the case of 5-residue loops in the correct environments, the average error was 0.6Å, as measured by local superposition of the loop mainchain atoms alone. For 8-residue loops in the correct environments, 90% of the loops had less than 2Å mainchain RMS error, with an average of less than 1.2Å.

Even 12-residue loops are modeled with useful accuracy in 30% of the cases. To simulate comparative modelling problems, the loop modelling procedure was evaluated by predicting loops of known structure in only approximately correct environments. Such environments were obtained by distorting the anchor regions, corresponding to the three residues at either end of the loop, and all the atoms within 10Å of the native loop conformation for up to 2-3Å by molecular dynamics simulations. When the RMSD distortion of the environment atoms is 2.5Å, the average loop prediction error increases by 180, 25 and 3% for 4, 8 and 12-residue loops, respectively.

It is not anymore too optimistic to expect useful models for loops as long as 12 residues, if the environment of the loop is at least approximately correct. It is possible to estimate whether or not a given loop prediction is correct, based on the structural variability of the independently derived lowest energy loop conformations. Typically, the loop prediction corresponds to the lowest energy conformation out of the 500 independent optimisations. The algorithm allows straightforward incorporation of additional spatial restraints, including those provided by template fragments, disulphide bonds, and ligand binding sites.

REFERENCES

Bowie JU, Luthy R, Eisenberg D., "A method to identify protein sequences that fold into a known three-dimensional structure", *Science*, 1991.

Debe DA, Danzer JF, Goddard WA, Poleksic A., "STRUCTFAST: Protein sequence remote homology detection and alignment using novel dynamic programming and profile-profile scoring", *Proteins*, 2006.

4

Biological Databases

Biological databases have become an important tool in assisting scientists to understand and explain a host of biological phenomena from the structure of biomolecules and their interaction, to the whole metabolism of organisms and to understanding the evolution of species. This knowledge helps facilitate the fight against diseases, assists in the development of medications and in discovering basic relationships amongst species in the history of life.

The biological knowledge is distributed amongst many different general and specialised databases. This sometimes makes it difficult to ensure the consistency of information. Biological databases cross-reference other databases with accession numbers as one way of linking their related knowledge together.

INFORMATION RETRIEVAL OF BIOLOGICAL DATABASES

Information retrieval in the biomedical research field is challenging. However, it is evolving rapidly and new techniques are being developed that may lead to more sophisticated information retrieval systems with both a high recall and a high precision.

In general, the goal of information extraction is to transform unstructured text data into structured information. Information extraction techniques have been used to support a wide range of applications such as the automatic extraction of protein interactions from the literature. They also have been used in the automatic determination of gene functions.

Name Entity Extraction

At the core of literature data mining is the identification of biological macromolecules (e.g., gene and protein names). Identification of macromolecule names is particularly challenging because different naming conventions can be used in different biological research areas. Also, a protein may be discovered through different research studies and given different names. An example of the latter case is protein p21. According to the information provided by the Atlas of Genetics and Cytogenetics in Oncology and Haematology database (http://www.infobiogen.fr/services/chromcancer/), it is also referred to as CDKN1A (cyclin-dependent kinase inhibitor 1A), WAF1 (Wild type p53 activated protein-1), CIP1 (CDK-interacting protein), SDI1 (Senescence cell-derived inhibitor-1), CAP20, and MDA-6 (melanoma differentiation-associated protein-1).

The approaches to name entity extraction can be divided into two categories: rule-based and dictionary-based. Rule-based approaches generate heuristic rules based on text features. The rules are generated either manually or by using machine learning techniques. They are then used for filtering documents and extracting name entities. Dictionary-based approaches consist of first constructing name entity dictionaries, and then detecting dictionary terms in the documents.

Rule-based approaches are particularly useful in identifying new names. This is essential in the biomedical domain because of the continuous discovery of new knowledge including, for example, the discovery of new genes and the discovery of new protein functions. However, if a macromolecule has multiple synonyms, rule-based approaches are not able to unify them. This problem can be solved by the dictionary-based approaches.

An example of a rule-based name entity extraction system detects two features of protein names: core terms and feature terms. Core terms consist of characteristic words containing uppercase letters, numerical figures and special symbols. Feature terms include words such as "receptor" that are frequently used to describe the function of a protein. This system first extracts core-terms from tokenised texts, then concatenates core-terms with feature-terms to construct noun-phrases that have no conjunctions and prepositions.

Based on the dependencies between core terms and feature terms included in the noun phrase, the system is capable of extracting material names with 94.7% precision and 98.84% recall. A rule-based approach has also been used in the identification of chemical names. In both of these cases the domain specific rules for protein name and chemical name identification were built manually.

Recently machine-learning techniques such as hidden markov models and support vector machines were used to support semi-automatic name detection. These techniques are used to train a system to learn the target rules from a set of sample pages. In, hidden markov models were trained with 100 MEDLINE (http://medlineplus.gov/) abstracts. Words in the abstract samples used during training were tagged according to their classes. A total of 11 classes, such as protein and DNA, were used.

The system achieved an F-score of 0.73, where the F-score is 2*Precision*Recall/(Precision+Recall). The quality of the result depends on the size and distribution of the training data. As shown in the classes that are highly represented in the training samples achieved better F-scores, whereas the classes that are poorly represented in the training data resulted in lower F-scores.

In general machine-learning approaches can generalise better (i.e., adapt to a new class of words) than systems that are based on manually created rules. The major disadvantage of these techniques is that they often require a large amount of tagged training samples whose creation can be time consuming.

Dictionary-based approaches consist of two steps: dictionary construction and dictionary-based name identification. The quality of the dictionary is critical to the quality of name identification. Dictionaries can be constructed manually, or semi-automatically. In a semi-automated method is provided to generate and curate a general-purpose protein name dictionary.

The dictionary is first constructed from the HUGO nomenclature database (http://www.gene.ucl.ac.uk/hugo/). Each entry contains the full name and alias of a gene symbol in HUGO. Corresponding synonyms are then extracted from OMIM, Swiss-Prot and TrEMBL (http://us.expasy.org/sprot/), and added to the dictionary. The curation of the dictionary includes an expansion phase, which expands known unambiguous acronyms within synonyms (e.g. IL1 can be expanded into Interleukin 1), and a pruning phase, which removes redundant, ambiguous and irrelevant synonyms.

After the dictionary is constructed, the next step consists of searching for dictionary terms against the literature. One of the possible approaches is to first parse abstracts into tokens. A set of candidate solutions (i.e., protein names that the token may represent) and two associated scores (boundary score and acceptance score) are assigned to each token. The boundary score in this case represents the probability of a mismatch, whereas the acceptance score represents the probability of a match.

A different approach of literature search translates both dictionary terms and scientific articles into an alphabet of DNA sequences by replacing each character with a unique, predetermined nucleotide combination. For example, "A" is replaced by "AAAC", and "E" is replaced by "AACG". The dictionary is, thus, converted into a sequence database where each record corresponds to one term in the original dictionary. The article is also converted into a continuous query string. Sequence alignment tools such as BLAST are then used to find the best matching dictionary sequences for the article. This approach has the disadvantage of not being able to recognise semantically similar terms that are syntactically very different (e.g., WAF1 and CIP1).

Document Indexing

Document indexing is the task of assigning terms to documents for retrieval purposes, and is mainly useful for the information management of libraries and databases. Currently, the most popular literature databases (e.g., Medline, PubMed) are provided by the National Library of Medicine. In order to make literature searchable, articles are indexed by hand with terms from the MeSH (Medical Subject Headings) (http://www.nlm.nih.gov/mesh) controlled vocabulary. An automatic indexing system was developed recently by the National Library of Medicine to overcome the increasing amount of effort associated with indexing and the lack of skilled indexers. This system first uses MetaMap Indexing, Trigram Phrase Matching, and PubMed Related Citations techniques to investigate citation titles and abstracts, and discover MeSH headings for them.

The combined list of headings produced by the three techniques is used as the index of the citation. MetaMap Indexing and Trigram Phrase Matching generate lists of UMLS (Unified Medical Language System) terms (http://www.nlm.nih.gov/mesh/umlsforelis.html) for each citation using different algorithms. MetaMap Indexing parses citation text into noun phrases, collects all UMLS terms containing one or more noun phrases or their variants, and ranks the candidate UMLS terms according to their similarities to all the noun phrases in the citation. Trigram Phrase Matching also parses text into phrases.

Then for each phrase, it computes its similarity with all the terms in UMLS, and records the most similar pairs. The pairs that appear most frequently in the citation text form the final UMLS term list. The UMLS term lists generated from the first and second techniques are mapped onto Mesh Headings. The third technique, PubMed Related Citations, first refers to the citations that have already been indexed, and finds among them the citations that are similar to the target citation. Indices (i.e., MeSH term lists) of the similar citations are then used to derive the MeSH term list of the target citation.

Document Clustering

Document clustering groups similar documents based on terms of interest they may have in common. These terms are in turn used to represent the features of the cluster. Currently controlled and curated vocabularies, such UMLS and MeSH, are widely used to express database queries and to support information extraction in the biomedical domain. However, because the majority and the most recent experimental results are stored by using a free-text format, document clustering can be used to generate more complete and up-to-date indices.

Document clustering can be used in concept discovery and the construction of terminology that facilitates index building. Although very useful, this type of technique still remains mostly unexplored. There are two applications of this technique.

The first application uses an unsupervised machine-learning algorithm to discover groups of related abstracts based on term co-occurrence. The system also decides on a list of terms that are used to represent each group of abstracts by comparing the frequency of usage of a given term in the group compared to its frequency of usage in all the abstracts. Terms whose frequencies of usage are higher within the group compared to the entire set of abstracts are candidates of choice for representing the group of abstracts.

The second application is based on a vector space representation of text. First, 26420 terms were extracted from the Gene Ontology to construct a term collection. Then, each document was represented as a k-dimensional vector, where k is the size of the term collection (26420 in this case). Each dimension in the vector corresponds to one term in the term collection, and the value of the vector entry represents the weight of the term in the document. The approach was used on a subset of annotations from the Swiss-Prot and Yeast databases associated with a 105 genes belonging to three biologically distinct functional groups.

The clustering algorithm is based on a fixed number of clusters where each cluster is represented by a single document. This document is called the medoid. Documents are then added to the clusters in such a way that the similarity between the medoids and the remaining documents in the cluster is maximised. Both of the above mentioned approaches generate clusters that are mutually exclusive. A soft clustering method, where documents can belong to more than once cluster.

Event Extraction

The objective of event extraction is to extract relations between genes, proteins, drugs or other biomaterials. Event extraction builds on name entity extraction, which was discussed earlier, by not only identifying names of entities but also detecting the relationships among these name entities.

The approach used by BioNLP is an extension of the rule-based approach which was also discussed in the Name

Entity Extraction section. BioNLP identifies sentences that contain protein names. It then searches for specific function words such as "inhibit" to establish the relationship between the protein names that are present within the sentence. The extraction rules in BioNLP are constructed manually. A successor system, PIES, integrates BioNLP with Kleisli, and Graphviz (a tool for drawing interaction pathways). PIES facilitates the construction of pathways by combining the processes of downloading abstracts, extracting protein-protein interaction information, and visualising pathway diagrams.

The extraction rules for events used in BioNLP are manually created. Similar to the case of name extraction, event extraction can also be performed by using machine-learning techniques that aim at deriving the rule automatically. For example, uses hidden markov models. It consists of identifying tokens of the input text, tagging the tokens (e.g., with "noun" or "verb"), grouping the words into phrases, grouping the phrases of interest, and finally storing the desired information in templates that may, for instance, include two protein names and the type of the relationship between them. This approach was experimentally tested on three relationships among proteins (interacts with, associates with, and binds to). Recall and precision were found to be as high as 63%, and 81%, respectively.

All of the event extraction tools discussed so far in this section focus on simple sentence constructs such as in "protein inhibits protein". However, a biomedical abstract can include complicated sentence structures. For example the sentence, "Interestingly, under the same condition, forskolin (20 mumol/L) stimulated the phosphorylation of LPS- and PNA-triggered p38 MAPK of murine peritoneal suppressor macrophages, suggesting that activation of p38 MAPK is regulated positively by both PKC and PKA."

Several event extraction systems have been developed to address this issue. These systems extract noun phrases from complex sentences in natural language text. Once these noun phrases are extracted, it is possible to use any of the techniques to continue the process of event extraction, since noun phrases have a very simple structure. Examples of systems that attempt to extract events from complex natural language text include, EDGAR, and GenScene. These systems rely on English language constructs to identify the precise relationship between target names (e.g., gene names or protein names) and target verbs. Target names and verbs can be extracted from various sources.

For example, both EDGAR and GeneScene use UMLS for this purpose. Additional and more specific target names and verbs can also be specified by an expert scientist for a given application. Very high precision can be achieved by these systems. For example, GeneScene achieves an average precision of 70% while the system achieves a precision of 80%. However, the recall rate can be low in these systems. For example, in the recall rate was 48%. Among the reasons for this low recall rate is the misclassification of complex sentence structures.

A different approach to event extraction uses techniques that have traditionally been used for literature indexing and document indexing. This approach measures statistically significant co-occurrences of genes in abstracts. The underlying assumption to this method is that genes that have a high level of co-occurrences must be interrelated. This approach was tested on abstracts that were filtered by using the "DNA repair" function and was shown to be able to extract functionally related genes. This approach can detect related entities even if the entity names occur in distant sentences.

An interesting study that compares the impact of the text unit used in information extraction. This study concluded that recall rates decrease and precision rates increase if the text unit used goes from abstract to sentence pairs to sentence to phrases. This indicates that a multi-tier approach to event extraction is needed. For example, one can start with an approach which operates on abstracts. Results obtained from this stage can then be refined by using an approach by EDGAR or GeneScene. This step can concentrate on the sentences identified in the previous step. It can use natural language processing techniques on these target sentences to identify complex language constructs and produce noun phrases. These noun phrases can then be processed by using techniques such as the one proposed in BioNLP.

GENOMIC INFORMATION RETRIEVAL

Sequence data such as DNA, RNA and amino acid sequences represents an important part of the available biomedical data. The number of complete eukaryote genome sequences available at NCBI has nearly doubled in the past year, to a total number of eight as of September 2002. Currently, human, mouse, rat, zebra fish, fruit fly, nematode, mosquito, baker's yeast, fission yeast, rice, oat, barley, thale cress, wheat, corn, soybean genomes and others are already available online at NCBI (http://www.ncbi.nlm.nih.gov/mapview).

In the post-genomic era, a substantial research effort is focusing on discovering genes from the genomic DNA sequences, and on identifying the functions of genes and proteins. Traditionally, genes and their functions were identified through experimental methods such as genetic linkage analysis, mutant study, and gene knockout. The function and structure information of genes or proteins from a given organism can be used to infer the function and structure of their homologous genes or proteins in another organism. Homologous genes are genes that share the same evolutionary history. Their sequences are usually similar and they can be found by using sequence alignment techniques.

For example, RNA polymerase sigma factor rpoD (Sigma-70) is a gene family that contains homologous genes in various organisms. All of them share similar DNA sequences, and their encoding proteins play similar functions in the initiation stage of the transcription from DNA to RNA. Thus, information related to a well-known gene in a given organism can be used to facilitate the study of its homologous genes in other organisms.

The large volume of genomic sequence data available presents an unprecedented opportunity for new discoveries through computational genome analysis. Homology searches are the building blocks of many studies, such as comparative genomics, gene prediction and phylogenetic analysis. In this section a brief introduction to sequence alignment is presented, followed by a description of the challenges facing homology based searches, and how information retrieval techniques can be used to improve sequence alignment-based homology searches.

Sequence Alignment

Sequence alignment of two sequences consists of superimposing the strings representing the two sequences in order to maximise the number of residues that are in common. Once the alignment is constructed, a residue in a given position that is common to both sequences indicates a common evolutionary origin, and this position is deemed to be conserved during evolution. Positional mismatches can indicate insertions, deletions and point mutations and they represent regions of the sequences that have not been conserved.

Dynamic programming approaches were initially used to find the best possible alignment between two sequences over their entire lengths. This type of alignment is called global alignment. Smith and Waterman proposed an algorithm that can identify maximally similar subsequences in two sequences. This algorithm is based on dynamic programming techniques. In this algorithm each deletion of length is assigned a weight . Also, a similarity score is assigned to each pair of residues.

The first sequence is denoted by and the second sequence is denoted by . The optimal homologous subsequence can then be found by locating the largest value of for and , and tracing back sequentially to determine the path that led to the entry. Many homologous sequences usually share several short homologous regions that are interleaved with residues that are completely different in the two sequences. These short homologous regions may appear in different orders in different sequences, and many of these regions may be repeated. This fact suggests that a local alignment may be more efficient than a global alignment.

Greedy algorithms have been used instead of dynamic programming in order to make the alignment computationally more efficient. However, the approach is only suitable for the comparison of two genome sequences that are highly similar.

Homology Search

Since scientists are often interested in finding homologous sequences to a given input (query) sequence from a large database, an algorithm based on dynamic programming techniques may be impractical because of the high space and time complexity. Even the simplest homology search requires the alignment of a query sequence against all the sequences in a database such as Genbank, which according to the NCBI website, contains more than 18 million sequences.

In order to make the homology searches against a large number of sequences efficient enough while remaining sensitive (i.e., being able to find distant homologous sequences), most homology search applications separate the process into two steps: a quick and inexpensive search to find a list of possible candidates from the entire list of sequences in a database followed by a computationally intensive alignment step for the reduced list of sequences generated in the first step. The local alignment techniques mentioned above are only used in the second step. In the first step, indexing techniques that are borrowed from information retrieval are used instead.

Based on the data structure used in the first step, the algorithms for homologous sequence search can be divided into two groups: hash-table-based and suffix-tree-based. Hash table-based approaches typically use several steps. In the first step the sequences in the database are indexed by using all possible sub-strings of a given size. These sub-strings, which are also called seeds, represent the hash key in the hash table. Each hash key points to the sequences in the database that contain the corresponding seed. This step is usually performed offline and the hash table needs to be updated only when the database instance changes (e.g., when more sequences are added to the database). The remaining steps are performed online for each query sequence (i.e., input sequence).

These steps consist of the selection of seeds from the hash table that match sub-strings in the query sequence; and the extension and combination of the selected seeds using local alignment algorithms. The seed combination and extension is needed in order to obtain longer homologous regions. The most important factor in the indexing is the size of the hash key (i.e., the size of the sub-string). This is also the size of the seed used during the matching process. Longer seeds increase the computational efficiency of the algorithm. However, longer seeds will also decrease the sensitivity of the algorithm in such a way that distant homologies may not be found.

For DNA sequence alignment, MEGABLAST uses a seed of size 28, and BL2SEQ(blastn) uses a seed of size 11. When a query sequence is aligned against the database, the sub-strings of the query sequences are matched against the seeds in the hash table. In most cases only exact matches of consecutive residues in a seed are considered. In some cases and in order to avoid the seed extension phase that can be computationally intensive, non-consecutive seed matches are used. PatternHunter is an example alignment algorithm that uses this approach. In PatternHunter, the residues of a hash table seed are interleaved with "don't care" filler residues in order to allow matches to sub-strings in the query sequences that do not include all the residues in the seed one after the other. This algorithm has been shown to be highly sensitive while being computationally efficient.

The seed extension stage in a homology search is computationally intensive. In this stage, an attempt is made to extend the seed in both directions in an effort to obtain longer homologous subsequences. There are some algorithms that can reduce the computational time involved in this stage by imposing restrictions on when the seed extension can be performed. For example, performs seed extension only when two seeds are found within a threshold distance of each other. This leads to fewer seeds being extended and therefore the computational time associated with this stage of the homology search is reduced.

Another technique, called X-Drop, was also used to reduce the complexity of the seed extension phase. This technique extends seeds in both directions until the similarity score achieved falls below the highest similarity score achieved so far by a threshold value.

This technique was used in conjunction with the previously mentioned techniques for reducing the computational cost of seed extension and it was also used in a greedy-based homology search algorithm. The homology search algorithms can either be hash-table-based or suffix-tree-based. A suffix of a given sequence is a subsequence that begins at any position in the sequence and includes all the residues until the end of the original sequence.

Suffix-tree-based algorithms convert sequences into suffix trees. A suffix tree stores all possible suffixes of an input sequence where each leaf node of the tree represents a unique suffix. This algorithm builds a suffix tree for each target sequence and then compares the query sequence to the suffix tree of the target sequence. This algorithm is very efficient and was used to compare sequences of genomes. Homology search algorithms are an active area of research that is driven by the availability of new entire genome sequences.

While similarity algorithms for small sequences have matured enough to be able to offer scientists a range of techniques from highly sensitive to computationally efficient, there is still a need for homology search algorithms that can align very large sequences efficiently. These algorithms are needed in order to facilitate comparative studies of two or more genomes. Phylogenetic studies can also benefit from the alignment of multiple genome sequences.

DATABASE INFORMATION RETRIEVAL

Along with the rapid development of web technologies, an increasing amount of biological information has been organised into databases and made available for public access through the web. The Journal of Nucleic Acids Research has been documenting the status of public biological databases in its first issue over the last seven years. According to this journal, there were only 200 databases in 1999. This number has grown to 386 by the end of 2002.

These databases not only include sequences, literatures, and experimental data (e.g., microarray data, 2D-gel graphs, structure coordinates), but also include additional information such as gene annotations, protein classifications, protein domain feature summaries, pathway graphs, genome map views, structure images derived from coordinates, microarray experiment annotations, etc. The information contained in these databases can facilitate the discovery of new biological knowledge.

Some of the popular databases include the online databases supported by NCBI (National Center of for Biotechnology Information) or EBI (European Bioinformatics Institute). NCBI is a comprehensive resource for biological data. The databases supported by NCBI include GenBank, OMIM, and molecular modeling database among others. The NCBI also provide search tools such as Entrez which links records in different databases and the Blast family of tools for sequence similarity searches. The EBI SRS server houses the SRS system which integrates both databases and sequence analysis applications.

SRS covers 400 different databases including Medline and the EMBL Nucleotide Sequence Database. Online databases have helped promote data availability. Prior to web databases, information was difficult to obtain since it mainly resided in journal archives. There are other characteristics of the biological data that are equally important to availability. These include quality, completeness and flexibility of access. The latter two characteristics are often equated to supporting interoperability among biological databases.

The quality and completeness of the data are usually hard to assess in most of the cases. For instance, the quality of structure data can be evaluated because data must adhere to specific rules, and thus, the correctness of structures based on their coordinate sets can be evaluated. However, assessing the quality of annotations is not trivial. Scientists may sometimes test the quality of the data in a new database by comparing the result of a query they are familiar with to the result they expect. However, this test may not be indicative of the quality of the remainder of the records in the database.

The large number of available databases has resulted in an active research that focuses on supporting interoperability among these databases. Support for the interoperability among databases addresses two of the above-mentioned factors, completeness and flexibility of access. If the databases are integrated then scientists can extract all of the available information from all the databases (completeness) through a single point of access (flexibility).

Entrez uses hypertext linking to relate DNA and protein sequence data and related information from various databases. Entrez uses two different types of links. Simple cross-references are used to; for example, link a protein sequence to its corresponding DNA sequence. Other more complex hypertext links are generated from similarities among a set of sequences or a set of abstracts. Thus, sets of abstract that are "neighbors" using a specified similarity measure are linked together.

The SRS system, which was briefly introduced before, integrates a large number of databases. This integration system has been shown to be scalable for a large number of databases in a production environment. SRS uses a sophisticated indexing mechanism to link related records from different databases. The indexing procedure used in SRS is similar to techniques used for literature indexing. In addition, SRS supports the use of views. Views allow scientists to create a customised representation of a subset of the data. Support for views can eliminate the need for specialised databases.

There are several challenges associated with the integration of multiple biological databases. Some of the major factors that can hinder the efficient integration of biological databases are data redundancy, data replication, and heterogeneous data representation. Data in a given database can be replicated totally (redundancy) or partially (replication) in another database.

The protein domain information and member protein alignment within each cluster are similar in these databases. For example, the zinc-containing alcohol dehydrogenases protein family information is provided in PROSITE (entry PS00059, PS00058), Pfam (entry PF00107), and BLOCKS (entry IPB002328). All of these databases provide member proteins information, usually through hyperlinks to protein sequence databases. InterPro is a database aiming to address the issue of data redundancy and to support the interoperability among databases in the case of protein classification.

Terminology Systems

Experimental biology generates large amounts of data from genome sequencing, microarray analysis, x-ray of macromolecular structures, etc. This information resides in the literature and in various autonomous databases. The rapid increase in the volume of literature data and the large number of databases render the task of searching for new findings difficult, in particular for a very specific domain. In order to support knowledge exchange among different life science disciplines, terminology systems, which provide common nomenclatures are critical.

Terminology systems are classified into three categories: controlled vocabularies, task-specific ontologies, and pure ontologies. Controlled vocabularies are task independent and do not support logical reasoning. Support for logical reasoning allows the ability of a given concept to be derived from one or more other concepts. Task-specific ontologies are restricted to specific applications and support logical reasoning. Pure ontologies can be re-used for various tasks and support logical reasoning.

Controlled vocabularies define a set of terms for a specific domain. They are often used for literature indexing or gene annotation. Well-known examples of controlled vocabularies include MBO, MeSH, GO, UMLS. All of these vocabularies were developed to facilitate communication of knowledge among experts. Controlled vocabularies are usually handcrafted and structured in hierarchical trees where a descendant node corresponds to a sub-class of the parent node. MBO covers biological objects (e.g., gene, protein), experimental procedures and computational aspects of molecular biology. MeSH is used by the National Library of Medicine to index and catalog biomedical literatures. This controlled vocabulary is organised in a hierarchical structure that consists of several levels of specificity for the concepts.

The Gene Ontology consortium (GO) is focused on developing structured vocabularies for molecular functions, biological processes, and cellular components that can be used across species. GO contains three independent hierarchies: one for molecular function, one for biological processes and one for cellular location. A simple tree-like inheritance structure is sufficient for controlled vocabularies and makes the implementation of a controlled vocabulary simpler and more interpretable by human experts.

UMLS integrates a collection of more than 60 vocabularies. Some of these vocabularies have multiple editions. Terms in these vocabularies are grouped in two levels. First, terms from different vocabularies that are lexical variants of each other, such as Zea mays, ZEA MAYS, Corn, corn, Indian Corn, Maize, maize, are clustered into a unique concept, and given a unique ID. Second, concepts are linked through inter-concept relationships (e.g., broader, like, parent, child, sibling, etc.). These inter-concept relationships are either inherited from the original source vocabularies or generated by UMLS. Many source vocabularies of UMLS place the concepts they include in some context (e.g., a hierarchical organisation). To preserve these various contextual views, UMLS does not merge the various source vocabularies into a single hierarchy. Thus, a single concept can appear in multiple hierarchies within UMLS. For example, "fruits" is a child of "technology, manufacturing, and agriculture", "food", and "Plant components" in different source vocabularies.

The second category of terminology systems, task-specific ontologies, can be further divided into ontologies for database schema description and ontologies for query translation. Ontologies have been used in the EcoCyc, HinCyc and GDB databases as database schema. Both the EcoCyc and the GDB ontologies use an object-oriented methodology where each concept is defined as an individual object class. Relationships are defined explicitly between classes and their attributes and between pairs of classes making the database schema more flexible, scalable and readable.

TAMBIS and BACIIS rely on task-specific ontologies for query translation. These systems integrate heterogeneous and distributed biological databases. The ontologies used in both of these systems are used to translate multi-database queries into database-specific sub-queries. The process includes two steps. User queries are decomposed into sub-queries that can be processed by a single source database. All the sub-queries are expressed by using ontology terms. Then sub-queries are translated into source database specific queries.

Pure ontologies are the third class of terminology systems. They are re-usable and application independent. Very few of this type of terminology system exist in the biomedical domain. For example, the GLAN CORE ontology (http://www.opengalen.org/) can be used for a variety of applications such as classification of surgical procedures, electronic healthcare records, clinical user interfaces, decision support systems, knowledge access systems, and natural language processing.

BIOLOGICAL DATABASE NORMALISATION

One of the major challenges for the integration of disparate data sources is normalisation; that is, determining whether two records refer to the sameunderlying entity. In the biological domain, the underlyingentity is determined by a nucleotide or amino acid sequence. Thus, considering sequence similarity should provide us witha way to find determine whether two database records cite to the same biological entity.

MIMI DATABASE

MiMI (Michigan Molecular Interactions) is a database consisting of molecules and the interactions between them. The vast majority of molecules are proteins. MiMI integrates interaction data from a variety of publicly available data sources, including DIP, BIND, GRID, HPRD, and others. The version of the MiMI database used comprises 66,713 molecule records and 138,454 interaction records. Each molecule in MiMI has at least one external identifier, which can be used to reference the molecule in other data sources.

So far as possible, interactions from different data sources are combined into a single MiMI entry based on the external database ID. Other than HPRD, the external IDs are not primary keys in the data sources MiMI integrates. They are instead references into databases whose primary function is to hold sequence data rather than interaction data. The most common identifiers are GIs assigned by NCBI and NCBI RefSeq identifiers. It is frequently the case that different interaction databases may refer to the same protein by non-overlapping sets of external IDs.

For example, HPRD asserts that a particular proteinhas RefSeq ID NP_ 000035, among other external identifiers, while BIND asserts that a particular protein has LocusLink ID 367. None of the external IDs provided by BIND or HPRD for this protein overlap, but LocusLink ID 367 also maps to NP_000035. Thus both the BIND and HPRD data should be merged to the same entry in MiMI, but are not. Thus, a first step at normalisation might be to collect lists of equivalent external IDs and then compute the connected components of the graph induced by this equivalence relation.

However, this does not entirely solve the problem. What should count as the same protein? Only proteins with exactly the same sequence? Only the same protein from a particular organism, but allowing for normal genetic variation? Homologous proteins from some group of organisms? The external ID mapping typically operates at the level of protein within a particular organism, but it can be the case that the same sequence is assigned multiple IDs even within the same database. Continuing with the androgen receptor example, the HPRD entry has a reference to GI 105325 and the BIND entry has a reference to GI 178628. These GIs have the exact same protein sequence, but for whatever reason, 105325 has been discontinued and been replaced -although by 113830, not 178628!

Thus, to a large extent the problem of duplicate entries in MiMI is a reflection of problems with the underlying data sources rather than flaws with the design of MiMI itself. However, one would like to compensate for these problems in as robust a way as possible. The solution is to allow searching MiMI based on sequence similarity. MiMI itself does not contain sequence data for each protein, so the external ID mapping is used to get the sequence data. It should be noted that simply finding similar proteins is not an ideal solution.

The true task is to find all the interactions in the database a given protein participates in, or more precisely to determine given two proteins A and B and the list of interactions of B, determine which of the interactions A also participates in. This is a somewhat more constrained problem than the general problem of determining if two given proteins interact. There is a great deal of ongoing work in the area of protein interaction prediction. The extended MiMI database simply returns the list of similar proteins and their interactions and leave it to the user to interpret the data.

BLAST

Given the set of sequences of proteins in the MiMI database and some sequence to query with, one needs some similarity metric. Ideally, this metric should be efficient to compute and should correlate well with homology. The BLAST (Basic Local Alignment Sequence Tool) algorithm serves this purpose. By default BLAST reports all alignments that are unlikely to have occurred by chance. However, we are interested in a much stronger notion. If we accept the assumption that homologous proteins are likely to participate in the same interactions, we would like all the proteins returned for a search to be homologous.

The score that BLAST reports for an alignment is the simply the sum of the score of each individual aligned residue pair. Identical and conserved pairs will contribute positively to the score; non-conserved pairs and gaps will contribute negatively. Thus the BLAST score will correlate strongly with the percentage of the query aligned, the percent identity, and the percent conserved. Additionally, the number of alignments E expected by chance with a given score for a given query and database is a simple function $E = kmne^{-\lambda S}$ of the score S, the size of the query and the database m and n, and constants of proportionality λ and k.

ATLAS DATABASE

The ATLAS database is an effort similar to MiMI that integrates various biological databases. It integrates several sequence and function databases in addition to the interaction databases, but its user interface is significantly less user-friendly and informative than MiMI's, and it suffers from the same problems as MiMI with respect to interaction data. There is no integration with BLAST or any BLAST-like tool.

NCBI's HomoloGene is a database of automatically detected homologous genes among 18 species. This data only covers a relatively small set of genes and is generally based on gene and not protein sequences, so although it could help in normalising the MiMI data it is not a complete solution. It also suffers from the same weaknesses as other static clusterings.

NCBI's BLink displays precomputed BLAST alignments for all proteins vs. all other proteins in NCBI's protein database. Ulysses, a very recent follow-on project to Atlas, uses a technique called interolog analysis to project interactions across species. Although the main focus is predicting in-teractions for human proteins based on homologous proteins in other organisms, this technique could be relevant to database normalisation. Ulysses uses the HomoloGene information to group homologous genes.

INTEGRATION OF INITIAL DATABASE

Sequence data and external ID mappings from several publicly available sequence databases were obtained as well as one database of interactions which also contained sequence data. These sequences and external IDs covered 63,280 out of 66,713 molecules in MiMI. The remaining 3433 records were primarily (2,220 records) molecules from HPRD that had no associated external IDs or sequence data. Some of these molecules were not in fact even proteins—for example there are molecules which report HPRD IDs of

"Sodium", "RNA" and even, mysteriously, "Mouse". There are also 1,155 LocusLink (EntrezGene) identifiers for which it was not possible to find associated proteins.

The existing MiMI database augmented with this additional sequence and identifier data is referred to as the extended MiMI database.

BLAST Parameters

BLAST has many tunable parameters which affect the ultimate score for each alignment. The extended MiMI database runs BLAST with the following parameters:

— -e 1e-15 — The expected number of times to see an alignment with this score in the database by chance, i.e. assuming amino acids are chosen randomly, should be very close to zero.

— -G 25 -E 2 — Gap opening and extension costs are chosen to be as expensive as possible. This is the most expensive recommended setting for protein-protein alignments.

— -A 10 — The BLAST algorithm works by aligning small initial windows and then expanding them. The current version of BLAST can find two separate windows in each possible hit and expand them simultaneously. Setting the -A parameter lower than the default of 40 forces the two windows to be closer together. For homologous sequences, one expects the sequences to be aligned more or less contiguously over most of their lengths. Thus, there is no reason to allow initial windows to be far apart.

— -M BLOSUM80—The -M parameter controls the score assigned to each aligned residue pair. The BLOSUM80 matrix was created by aligning protein blocks that have at least 80% similarity to each other. This matrix is most stringent matrix provided in the BLAST distribution.

Database Searching

The extended MiMI database can be searched in three ways: by external ID, by MiMI ID, and by sequence. It gathers the set of distinct protein sequences for the external ID or for all external IDs mapping to the MiMI ID, or the single provided sequence, and uses the above BLAST parameters to search for similar sequences. The returned sequences are grouped by the MiMI ID or IDs they refer to, and sort by BLAST score. If there is more than one matching sequence for a given MiMI ID, the highest score of any of the matches is used. The user can also sort by other criteria: percent of query aligned, E-value, percent identity, percent conserved. The iterative nature of interactive search means that the user can choose to re-fine the search with more appropriate criteria, although of course the hits should be returned in the most useful order by default.

For each result (MiMI molecule) the following is displayed: a graphical overview of each alignment with a tooltip with the external IDs for the hit sequence, a list of all external IDs for the MiMI molecule, a list of names for the MiMI molecule, and a list of the interactions the MiMI molecule participates in. The list of external IDs and names consists not only of those in the original MiMI database but also those external IDs and names found through the connected components. Each interacting molecule has a link so that the extended MiMI database can immediately be searched for the interacting molecule, and each external ID has a link to search the external database the ID came from. edu/mimi

The original version of MiMI contains additional information such as provenance and confidence scores for each interaction. This additional information is not replicated in the results page, since ideally there would be a link to the original MiMI database for this information. However, one weakness of MiMI is that its internal identifiers do not remain consistent between releases of the database. This on-the-fly search is superior to clustering proteins in the MiMI database based on sequence similarity, e.g. a HomoloGene-style approach. A clustering is by defi-nition a static object. Grouping MiMI proteins into static clusters and reporting interactions for the entire cluster would lose information relative to ranking molecules by their similarity to a given search.

When searching the extended MiMI database, BLAST is run against the set of distinct sequences which map to some MiMI molecule. This is 65,274 sequences comprising 66,513,740 residues. It takes approximately 4.5 seconds to search for a query sequence of 377 residues on a dual 2.0GHz Pentium IV Xeon with 3 gigabytes of RAM running Linux. It would be possible to precompute all the alignments; however, the storage space and computational requirements would be unreasonable given the short time to takes simply to run BLAST on the fly. However, if the MiMI database became an order of magnitude larger, precomputation on a large parallel cluster might become more attractive.

BLAST is fundamentally very easy to parallelize. The database can be partitioned horizontally and each query run against each partition, or the database can be replicated and the queries partitioned. The first solution can actually achieve a superlinear speedup—for each query, the database must be scanned, and if each partition fits in memory but the entire database does not then tremendous time savings can be achieved.

DATABASE ASSESSMENT

Now that we have a way of returning additional interactions for any given molecule, we should consider ways to evaluate the results as well as the ranking of results. In this sense the search is more akin to the ranked list of results returned by a Web search engine than the set of results returned by a database query satisfying some Boolean predicate. In the database sense, a result is simply correct or incorrect. In the information retrieval sense, some results are more relevant than others, and we would like for the most relevant results to be ranked first.

If we have the evaluation results on some development set, we can attempt to develop a reranking strategy that weights the various features of each result (BLAST score, E value, percent of query aligned, percent of hit aligned, percent identity, percent conserved, number of gaps, etc) so that the final result ranking is more appropriate. For each of the feasible evaluation strategies described below, informal experiments with a small set of common molecules were conducted to see if the evaluation strategy warranted further investigation. This development test set consisted of androgen receptor (AR), estrogen receptor (ER), tumor necrosis factor α (TNFα), and alcohol dehydrogenase.

Protein and Molecules are Homologous

One simple way to evaluate the results is simply to check whether each result is homologous to the original query. This is unsatisfactory for several reasons. First, we are not directly interested in homology—we are interested in interactions. Proteins may be homologous but only similar along part of their sequences, and thus unlikely to participate in many of the same interactions. In some cases it may not be clear whether we want to return a homologous molecule or not. For example, androgen receptor and progesterone receptor are both steroid receptors and share several domains, participate in many of the same interactions, yet are clearly not the same protein. So if we searched for an androgen receptor, we should first return all androgen receptors, then progesterone receptors, then perhaps other steroid receptors.

Simply considering whether or not molecules are homologous does not capture this kind of situation. Also, although it is not difficult to scan result sets and determine which are homologous to the query to determine the percentage of results returned that are correct (that is, precision) it would be quite expensive to examine the entire database by hand and determine which proteins are homologous to the query to determine recall, that is, the percentage of correct results in the entire database that were returned. In fact, the only feasible way to determine this is to use the same tool (BLAST) with less stringent parameters to search the database for possibly homologous proteins. Also, all of the proteins returned using the stringent BLAST parameters above should be homologous, that is, one would expect virtually 100% precision. This is confirmed on the development test set. So, this evaluation strategy is not satisfactory.

Accuracy and Recall of Interactions

Another way to evaluate the results might be to evaluate precision and recall of the list of returned interactions. That is, collect the set of interactions for all returned molecules, and check both how many are true interactions for the query molecule, and how many of the true interactions for the query molecule are on the returned list. This is a more attractive evaluation strategy than precision and recall of homologous molecules, but is much more difficult to evaluate. For each pair of molecules, the only definitive way to determine if they truly interact is via experiment. Even if there was a list of interactions separate from the databases from which MiMI was constructed, it may be the case that there are true interactions not on this list either.

Also, interactions in the original source databases for MiMI may not always be correct, and the source databases may be missing interactions in the comprehensive list. If we used this evaluation strategy, we would need to compensate for errors in the original data. Thus to correctly evaluate precision and recall of interactions, compensating for errors in the original data, we need complete lists of all proteins the query and all hits interact with. If it were possible to obtain this data on a scale large enough for a reasonable evaluation, then we would not need MiMI or extended MiMI. Thus, evaluating precision and recall of interactions, while attractive in theory, is not feasible in practice.

DATA MINING TECHNIQUES

Data mining techniques are the result of a long process of research and product development. This evolution began when business data was first stored on computers, continued with improvements in data access, and more recently, generated technologies that allow users to navigate through their data in real time. Data mining takes this evolutionary process beyond retrospective data access and navigation to prospective and proactive information delivery. Data mining is ready for application in the business community because it is supported by three technologies:

1. Massive data collection
2. Powerful multiprocessor computers
3. Data mining algorithms

Dynamic Data Access

Dynamic data access is critical for drill-through in data navigation applications, and the ability to store large databases is critical to data mining. The core components of data mining technology have been under development for decades, in research areas such as statistics, artificial intelligence, and machine learning. Today, the maturity of these techniques, coupled with high-performance relational database engines and broad data integration efforts, make these technologies practical for current data warehouse environments.

Data mining derives its name from the similarities between searching for valuable business information in a large database—for example, finding linked products in gigabytes of store scanner data—and mining a mountain for a vein of valuable ore. Both processes require either sifting through an immense amount of material, or intelligently probing it to find exactly where the value resides. Given databases of sufficient size and quality, data mining technology can generate new business opportunities by providing these capabilities:

— Automated prediction of trends and behaviours. Data mining automates the process of finding predictive information in large databases. Questions that traditionally required extensive hands-on analysis can now be answered directly from the data—quickly. A typical example of a predictive problem is targeted marketing. Data mining uses data on past promotional mailings to identify the targets most likely to maximise return on investment in future mailings. Other predictive problems include forecasting bankruptcy and other forms of default, and identifying segments of a population likely to respond similarly to given events.

— Automated discovery of previously unknown patterns. Data mining tools sweep through databases and identify previously hidden patterns in one step. An example of pattern discovery is the analysis of retail sales data to identify seemingly unrelated products that are often purchased together. Other pattern discovery problems include detecting fraudulent credit card transactions and identifying anomalous data that could represent data entry keying errors.

Data mining techniques can yield the benefits of automation on existing software and hardware platforms, and can be implemented on new systems as existing platforms are upgraded and new products developed. When data mining tools are implemented on high performance parallel processing systems, they can analyse massive databases in minutes. Faster processing means that users can automatically experiment with more models to understand complex data. High speed makes it practical for users to analyse huge quantities of data. Larger databases, in turn, yield improved predictions. Databases can be larger in both depth and breadth:

— *More columns*: Analysts must often limit the number of variables they examine when doing hands-on analysis due to time constraints. Yet variables that are discarded because they seem unimportant may carry information about unknown patterns. High performance data mining allows users to explore the full depth of a database, without preselecting a subset of variables.

— *More rows*: Larger samples yield lower estimation errors and variance, and allow users to make inferences about small but important segments of a population.

The most commonly used techniques in data mining are:

— *Artificial neural networks*: Non-linear predictive models that learn through training and resemble biological neural networks in structure.

— *Decision trees*: Tree-shaped structures that represent sets of decisions. These decisions generate rules for the classification of a dataset. Specific decision tree methods include Classification and Regression Trees (CART) and Chi Square Automatic Interaction Detection (CHAID).

— *Genetic algorithms*: Optimisation techniques that use processes such as genetic combination, mutation, and natural selection in a design based on the concepts of evolution.

— *Nearest neighbour method*: A technique that classifies each record in a dataset based on a combination of the classes of the k record(s) most similar to it in a historical dataset (where k ^{3} 1). Sometimes called the k-nearest neighbour technique.

— *Rule induction*: The extraction of useful if-then rules from data based on statistical significance.

Many of these technologies have been in use for more than a decade in specialised analysis tools that work with relatively small volumes of data. These capabilities are now evolving to integrate directly with industry-standard data warehouse and OLAP platforms.

Working of Data Mining

How exactly is data mining able to tell you important things that you didn't know or what is going to happen next? The technique that is used to perform these feats in data mining is called modelling. Modelling is simply the act of building a model in one situation where you know the answer and then applying it to another situation that you don't. For instance, if you were looking for a sunken Spanish galleon on the high seas the first thing you might do is to research the times when Spanish treasure had been found by others in the past. You might note that these ships often tend to be found off the coast of Bermuda and that there are certain characteristics to the ocean currents, and certain routes that have likely been taken by the ship's captains in that era. You note these similarities and build a model that includes the characteristics that are common to the locations of these sunken treasures.

With these models in hand you sail off looking for treasure where your model indicates it most likely might be given a similar situation in the past. Hopefully, if you've got a good model, you find your treasure. This act of model building is thus something that people have been doing for a long time, certainly before the advent of computers or data mining technology. What happens on computers, however, is not much different than the way people build models. Computers are loaded up with lots of information about a variety of situations where an answer is known and then the data mining software on the computer must run through that data and distill the characteristics of the data that should go into the model.

Once the model is built it can then be used in similar situations where you don't know the answer. For example, say that you are the director of marketing for a telecommunications company and you'd like to acquire some new long distance phone customers. You could just randomly go out and mail coupons to the general population—just as you could randomly sail the seas looking for sunken treasure. In neither case would you achieve the results you desired and of course you have the opportunity to do much better than random—you could use your business experience stored in your database to build a model.

As the marketing director you have access to a lot of information about all of your customers: their age, sex, credit history and long distance calling usage. The good news is that you also have a lot of information about your prospective customers: their age, sex, credit history etc. Your problem is that you don't know the long distance calling usage of these prospects (since they are most likely now customers of your competition). You'd like to concentrate on those prospects who have large amounts of long distance usage. You can accomplish this by building a model.

Applications of Data Mining

A wide range of companies has deployed successful applications of data mining. While early adopters of this technology have tended to be in information-intensive industries such as financial services and direct mail marketing, the technology is applicable to any company looking to leverage a large data warehouse to better manage their customer relationships. Two critical factors for success with data mining are: a large, wellintegrated data warehouse and a well-defined understanding of the business process within which data mining is to be applied such as customer prospecting, retention, campaign management, and so on. Some successful application areas include:

— A pharmaceutical company can analyse its recent sales force activity and their results to improve targeting of high-value physicians and determine which marketing activities will have the greatest impact in the next few months. The data needs to include competitor market activity as well as information about the local health care systems. The results can be distributed to the sales force via a wide-area network that enables the representatives to review the recommendations from the perspective of the key attributes in the decision process. The ongoing, dynamic analysis of the data warehouse allows best practices from throughout the organisation to be applied in specific sales situations.

— A credit card company can leverage its vast warehouse of customer transaction data to identify customers most likely to be interested in a new credit product. Using a small test mailing, the attributes of customers with an affinity for the product can be identified. Recent projects have indicated more than a 20-fold decrease in costs for targeted mailing campaigns over conventional approaches.

— A diversified transportation company with a large direct sales force can apply data mining to identify the best prospects for its services. Using data mining to analyse its own customer experience, this company can build a unique segmentation identifying the attributes of high-value prospects. Applying this segmentation to a general business database such as those provided by Dun & Bradstreet can yield a prioritised list of prospects by region.

— A large consumer package goods company can apply data mining to improve its sales process to retailers. Data from consumer panels, shipments, and competitor activity can be applied to understand the reasons for brand and store switching. Through this analysis, the manufacturer can select promotional strategies that best reach their target customer segments.

Each of these examples has a clear common ground. They leverage the knowledge about customers implicit in a data warehouse to reduce costs and improve the value of customer relationships. These organisations can now focus their efforts on the most important (profitable) customers and prospects, and design targeted marketing strategies to best reach them.

Decision Trees

Decision trees are data mining technology. As its name implies, It can be viewed as a tree. Decision tree algorithms tend to automate the entire process of hypothesis generation and then validation much more completely and in a much more integrated way than any other data mining techniques. They are also particularly adept at handling raw data with little or no pre-processing. Because decision trees score so highly on so many of the critical features of data mining they can be used in a wide variety of business problems for both exploration and for prediction.

They have been used for problems ranging from credit card attrition prediction to time series prediction of the exchange rate of different international currencies. There are also some problems where decision trees will not do as well. Some very simple problems where the prediction is just a simple multiple of the predictor can be solved much more quickly and easily by linear regression. Usually the models to be built and the interactions to be detected are much more complex in real world problems and this is where decision trees excel.

Specifically each branch of the tree is a classification question and the leaves of the tree are partitions of the dataset with their classification. For instance if we were going to classify customers who churn in the Cellular Telephone Industry a decision tree might look something like that found in Figure 1. You may notice some interesting things about the tree:

— It divides up the data on each branch point without losing any of the data (the number of total records in a given parent node is equal to the sum of the records contained in its two children).

— The number of churners and non-churners is conserved as you move up or down the tree.

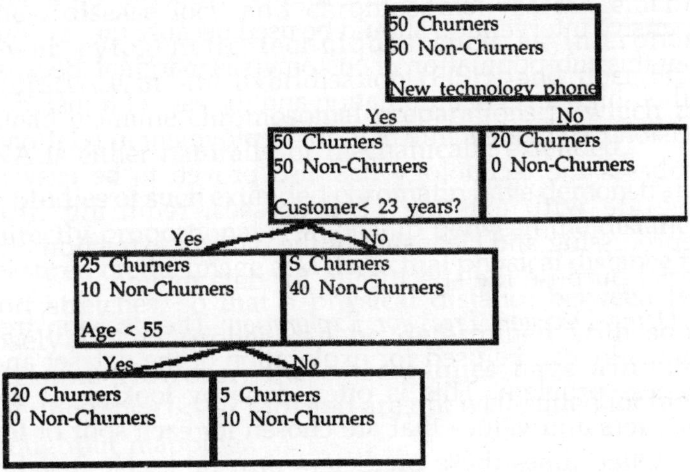

Figure 1. A decision tree is a predictive model that makes a prediction on the basis of a series of decision much like the game of 20 questions.

— It is pretty easy to understand how the model is being built (in contrast to the models from neural networks or from standard statistics).

— It would also be pretty easy to use this model if you actually had to target those customers that are likely to churn with a targeted marketing offer.

— You may also build some intuitions about your customer base. E.g. "customers who have been with you for a couple of years and have up to date cellular phones are pretty loyal".

Application of Decision Trees

Because of their tree structure and ability to easily generate rules decision trees are the favoured technique for building understandable models. Because of this clarity they also allow for more complex profit and ROI models to be added easily in on top of the predictive model. For instance once a customer population is found with high predicted likelihood to attrite a variety of cost models can be used to see if an expensive marketing intervention should be used because the customers are highly valuable or a less expensive intervention should be used because the revenue from this sub-population of customers is marginal. Because of their high level of automation and the ease of translating decision tree models into SQL for deployment in relational databases the technology has also proven to be easy to integrate with existing IT processes, requiring little preprocessing and cleansing of the data, or extraction of a special purpose file specifically for data mining.

Using Decision Trees for Exploration: The decision tree technology can be used for exploration of the dataset and business problem. This is often done by looking at the predictors and values that are chosen for each split of the tree. Often times these predictors provide usable insights or propose questions that need to be answered. For instance if you ran across the following in your database for cellular phone churn you might seriously wonder about the way your telesales operators were making their calls and maybe change the way that they are compensated: IF customer lifetime < 1.1 years AND sales channel = telesales THEN chance of churn is 65%.

Using Decision Trees for Data Preprocessing: Another way that the decision tree technology has been used is for preprocessing data for other prediction algorithms. Because the algorithm is fairly robust with respect to a variety of predictor types (e.g. number, categorical etc.) and because it can be run relatively quickly decision trees can be used on the first pass of a data mining run to create a subset of possibly useful predictors that can then be fed into neural networks, nearest neighbour and normal statistical routines—which can take a considerable amount of time to run if there are large numbers of possible predictors to be used in the model.

Decision Tress for Prediction: Although some forms of decision trees were initially developed as exploratory tools to refine and preprocess data for·more standard statistical techniques like logistic regression. They have also been used and more increasingly often being used for prediction. This is interesting because many statisticians will still use decision trees for exploratory analysis effectively building a predictive model as a by product but then ignore the predictive model in favor of techniques that they are most comfortable with. Sometimes veteran analysts will do this even excluding the predictive model when it is superior to that produced by other techniques. With a host of new products and skilled users now appearing this tendency to use decision trees only for exploration now seems to be changing.

The First Step is Growing the Tree: The first step in the process is that of growing the tree. Specifically the algorithm seeks to create a tree that works as perfectly as possible on all the data that is available. Most of the time it is not possible to have the algorithm work perfectly. There is always noise in the database to some degree (there are variables that are not being collected that have an impact on the target you are trying to predict). The name of the game in growing the tree is in finding the best possible question to ask at each branch point of the tree. At the bottom of the tree you will come up with nodes that you would like to be all of one type or the other.

REFERENCES

Minoru Kanehisa. *Post-Genome Informatics*. Oxford University Press. 2000.

Nei, M., *Molecular Evolutionary Genetics*. Columbia University Press, Irvington, New York. 1989.

Schulze Kremer S. (Editor), S. Schulze-Kremer (Editor) *Advances in Molecular Bioinformatics*. IOS Press. 1994

Steen Knudsen. *A Biologist's Guide to Analysis of DNA Microarray Data*. Wiley-Liss. 2002

5

Analysis of Gene Expression

Gene expression is the process in which the inheritable information in a gene, such as the DNA sequence, is made into a functional gene product, such as protein or RNA.

Several steps in the gene expression process may be modulated, including the transcription step and the post-translational modification of a protein. Gene regulation gives the cell control over structure and function, and is the basis for cellular differentiation, morphogenesis and the versatility and adaptability of any organism. Gene regulation may also serve as a substrate for evolutionary change, since control of the timing, location, and amount of gene expression can have a profound effect on the functions (actions) of the gene in the organism.

DNA ARRAYS AND GENE EXPRESSION

DNA microarrays are tools for exploration and discovery on a genomic scale. Microarrays are extensions of hybridisation-based methods which have been used for decades to identify and quantitate nucleic acids in biological samples: samples of interest are labelled and allowed to hybridise to the array; after sufficient time for hybridisation and following appropriate washing steps, an image of the array is acquired and the representation of individual nucleic acid species in the sample is reflected by the amount of hybridisation to complementary DNAs immobilised in known positions on the array. The idea of using ordered arrays of DNAs to perform parallel hybridisation studies is not in itself new; arrays on porous membranes have been in use for years.

However, many parallel advances have occurred to transform these rather clumsy membranes into much more useful and efficient methods for performing parallel genetic analyses.

First, large-scale sequencing projects have produced information and resources that make it possible to assemble collections of DNAs that correspond to all, or a large fraction of, the genes in many organisms from bacteria to humans.

Second, technical advances have made it possible to generate arrays with very high densities of DNAs, allowing for tens of thousands of genes to be represented in areas smaller than standard glass microscope slides.

Finally, advances in fluorescent labelling of nucleic acids and fluorescent detection have made the use of these arrays simpler, safer and more accurate.

In standard spotting applications large collections of DNA samples are assembled in 96 or 384 well plates. A robot positions a cluster of specially designed tips into adjacent sample wells, draws up approximately a microliter of DNA, and deposits a small spot of each onto coated glass microscope slides. By printing successive DNA samples in a staggered fashion, thousands of spots can be printed in an area corresponding to a cluster of 4 wells of a 96-well plate or 16 wells of a 384-well plate.

Typically 100-200 copies of a given array can be printed simultaneously by successively touching the cluster of printing tips to each slide before washing and drying the tips and reloading with the next set of DNA samples. After a complete set of samples is spotted, the slides are treated to attach the DNA stably to the surface and to minimise nonspecific binding of probes to the slide. DNA microarrays have been used for a variety of purposes; essentially any property of a DNA sequence that can be made experimentally to result in differential recovery of that sequence can be assayed for thousands of sequences at once by DNA microarray hybridisation.

Genome Expression Monitoring

The ultimate technical goal of whole genome expression monitoring is to be able to determine to absolute representation of every RNA species in any cell or tissue sample of interest. One would like to be able to isolate RNA from a sample, make a labelled representation of the isolated RNA, hybridise this labelled probe to a microarray, and read out the amount of each RNA species in the original sample. However, there is a complex relationship between the amount of input RNA for a given gene and the intensity of the probe signal at a corresponding hybridisation target. This relationship depends on a multitude of factors, including the labelling method, hybridisation conditions, target features and the sequence of the gene.

Therefore, microarray based methods are best used to assay the relative representation of RNA species in two or more samples. Fortunately, differences in gene expression between samples—i.e. where and when it is or is not expressed, and how it changes with perturbations, are what matters most about a gene's expression. Knowing the absolute abundance of the RNA provides only marginally further utility.

Two-colour hybridisations: To maximise the reliability and precision with which we can quantitate differences in the abundance of each RNA species, we directly compare two samples by labelling them with spectrally distinct fluorescent dyes and mixing the two probes for simultaneous hybridisation to one array. The relative representation of a gene in the two samples is assayed by measuring the ratio of the fluorescence intensities of the two dyes at the cognate target element.

Experimental questions about gene expression to which DNA microarrays can be applied: The simplest kind of gene expression survey compares the relative abundance of mRNA corresponding to each arrayed gene, between two different cell or tissue samples. For example, an experiment might compare cells before and after an experimental perturbation, or at successive times during a temporally staged process, or between stages of differentiation, or one might compare the pattern of mRNA's expressed in a mutant cell with that of its wild-type counterpart.

However, the flexibility of this experimental approach allows much more precise information about the regulation of each gene's expression to be collected. For example, translation rates can be measured for each gene by comparative hybridisation of polysomal RNA and total mRNA to a microarray. The distribution of each polyadenylated RNA between the nucleus and cytoplasm can be assessed by differential hybridisation of nuclear and cytoplasmic polyadenlyated RNA samples. A similar approach can allow other kinds of subcellular localisation to be surveyed for the entire set of arrayed genes.

Type-I and Type-II experiments: Many interesting biological questions can be studied using the simplest form of two-colour hybridisation experiments, where two samples are directly compared on a single array. However, such direct comparison is impractical for complex questions requiring the comparison of multiple samples. An alternative experimental design using a common reference sample in each of the hybridisations required for an experiment. This maintains the essential internal-control aspect of two-colour hybridisation, but allows for inferred comparisons to be made among large numbers of samples without requiring that every pairwise comparison be performed. The most common use of this method has been in time-course experiments where each timepoint is compared to an initial timepoint.

Since the amount of a given gene at each timepoint is measured relative to a fixed reference, the behaviour of each gene across the timecourse can be studied. The reference sample need not be related to the samples being examined. The most important attribute of the reference sample is that it should provide a hybridisation signal, and thus a non-zero denominator for the hybridisation ratio, at each target element on the array. For any large collection of samples to be compared, a convenient approximation of this ideal reference sample is an equal mixture of material from each of the samples. In such a mixture, any gene represented in one of the samples will be represented in the reference.

Making and using printed DNA microarrays requires two pieces of hardware: a robot to produce the microarrays, and a device to image the hybridised arrays. The excitement over microarray technology in large research labs and the biotech industry has been tempered elsewhere by concerns over the cost and expertise required to perform microarray experiments.

DNA Preparation

Choice of target: For large-scale gene expression studies, a specific hybridisation target is required for every gene to be analysed. As essentially any double-stranded DNA sample can be printed onto a treated glass surface, the choice of target is largely dictated by the resources available for obtaining representations of the genes to be studied. For organisms whose genomes have been fully sequenced, the most straightforward approach is to amplify every known and predicted open reading frame in the genome using PCR.

For unsequenced or partially sequenced genomes, or for genomes with large and numerous introns, this strategy is not practical. For many organisms, such as humans and mouse, extensive resources have been directed at determining partial sequences of clones from cDNA libraries. These expressed sequence tags (ESTs) can be used to identify distinct mRNA transcripts, and individual cDNA clones corresponding to each of these transcripts can be used as the source of gene-specific targets in an array. For unsequenced organisms without a genome sequencing project, or where no sequence is yet available, clones from cDNA libraries can be used as targets. Expression data can be collected for each of these clones without any prior information about the sequence, map position or identity of the gene.

It is likely that some benefits could be found in using DNA fragments smaller than full open reading frames or 1000 basepair cDNA's to represent each gene. For example, with yeast genes having significant homology to each other, PCR products representing the most divergent parts of these genes could be used in place of the entire coding region to improve discrimination between the two genes. The optimal representation of each gene is a complicated problem currently under investigation.

PCR Amplification: In general, a 100µl PCR reaction provides sufficient DNA for printing up to 1000 arrays. The actual parameters of the amplification will depend on the type of template and primers being used. It is important, however, to limit the components of the PCR to template, primers, nucleotides, magnesium, standard buffer and enzyme. Common additives such as glycerol or gelatin can interfere with the printing process by altering the surface tension of the drop or competing for surface attachment sites on the slide, and should not be used.

Preparing DNA for Printing: To clean up PCR products and prepare the DNA for printing, isopropanol-precipitate the PCR products and resuspend the DNA at approximately 100 ng/µl in 3x SSC. Precipitation can be carried out either in the thermocycler plates used for PCR or in V-bottom 96-well tissue culture plates. Start by adding, to each well, 10 µl of 3M sodium acetate, pH 5.2, followed by 110 µl isopropanol. Place the plates at -20C for one hour. Precipitate the DNA by centrifugation at 4C at 3500g for 1 hour.

Place two or three folded paper towels beneath the plate to prevent cracking. Stacks of up to four V-bottom plates can be spun in one adapter provided that a cushion of paper towels is placed between each plate and that the plates are carefully taped together to prevent slippage. After precipitation, carefully remove the liquid by aspiration or by inverting the plate. Wash the pellets with 100 µl 70% ethanol and spin again for 30 minutes. Aspirate or decant the liquid again, and dry the pellets in a rotary evaporator fitted to accept 96-well plates. Resuspend the pellets in 10-15 µl 3x SSC (pH 7.0) at 4C for at least 12 hours; care should be taken to seal the plates so that little evaporation occurs, as excessive salt or DNA concentration can interfere with printing.

After resuspension, transfer the solution to soft-bottom plates for printing. Seal the plates and store at 4C or -20C until use. Immediately prior to printing, the plates should be spun lightly to ensure that all material is at the bottom of the well. Some users have had difficulty obtaining consistent results using this precipitation process. Many additional methods are available for recovering purified DNA from PCR reactions, and numerous kits for cleaning up PCR reactions can be purchased.

The basic goal of this step is for the final DNA solution to be as free of contamination as possible; salts, detergents, glycerol, gelatin and particulate matter such as resins from columns can interfere at numerous later stages in the printing process. If these are used, extreme care must be taken to ensure that they are removed from the final DNA solution. Any alternative method for preparing DNA solutions for printing should be tested on a small scale, and taken through the entire printing and hybridisation process to ensure that it is compatible. An alternate method that has been used successfully is described below.

Sephacryl purification: This method is based on using a simple gel filtration resin that retains salts and other small molecules while excluding PCR products. As the DNA elutes in water, it can simply be dried down, eliminating the need for precipitation. As mentioned above, great care must be taken to prevent the resin from contaminating the printing solutions. To purify 100 µl PCR reactions, prepare Sephacryl S400 resin by mixing well and diluting 2:1 in double-distilled H_2O (ddH_2O). Aliquot 100 µl resin into 96-well filter bottom plates. Wash four times by adding 400 µl ddH_2O to plates and aspirating using a vacuum manifold. Nest these plates in polypropylene 96-well plates and spin for five minutes at 4C at 2000 rpm to remove all remaining liquid. Add 100 µl ddH_2O and spin as above to test recovery. Add 100 µl PCR product and spin into clean plates. Dry down in rotary evaporator and resuspend in 3x SSC as above.

Slides Preparation

To prepare poly-lysine coated slides, place standard glass microscope slides in metal or glass racks so that the slides are positioned vertically, and place the rack in an appropriately sized glass chamber. Make sure that you use slide racks that will fit in an available low-speed centrifuge as they must be spun dry at the end of the coating process, and it is best to avoid having to transfer the slides from one type of rack to another. Although most slides come in boxes labelled "pre-cleaned", it is essential that they be stripped clean prior to coating. Prepare 500 ml alkaline wash solution by dissolving 50 g sodium hydroxide in 200 ml

ddH_2O and adding 300 ml 95% ethanol. Completely submerge the slides in this solution for at least two hours. After cleaning, extensively rinse the slides by five cycles of adding clean ddH_2O, rocking for five minutes and rinsing.

Keep the slides submerged in ddH_2O; once the slides are washed, it is essential that they be exposed to air as little as possible, as dust particles will interfere both with the coating and printing processes. Prepare a coating solution containing 35 ml poly-L-lysine, 35 ml sterile-filtered PBS, and 280 ml ddH_2O. Pour this solution into a clean glass chamber, and quickly transfer the slides from the final water wash into the coating solution. Rock the solution gently for one hour. Dunk the coated slides in a chamber containing clean ddH_2O and plunge up and down five times. Transfer the slide rack to a low-speed centrifuge and spin the slides dry. Wrap the dry slides in aluminum foil to keep out dust. Place the wrapped slide rack in a vacuum drying oven at 45C for 10 minutes with the vacuum on.

Remove the slides from the foil and place in a clean plastic slide rack. Store at room temperature with the slide box tightly sealed. The slides are not immediately ready for printing, but must be "cured" for at least a few weeks to allow for the surface to become sufficiently hydrophobic. The hydrophobicity of the coated slide surface is important in maintaining the small size of the printed DNA spots necessary for high-density arraying. A simple test for the readiness of a batch of slides is to take one slide and place 150 µl H_2O on the surface. Turn the slide 45 degrees to the horizontal and watch as the drop moves down the slide. If it leaves no noticeable trail, the slides are probably ready for use. However, the only reliable way to tell if a batch of slides is ready for printing is to print a test array on a few slides from all available batches.

Printing DNA Microarrays

The arraying robot used in printing DNA microarrays is a

variation of a standard "pick-and-place" robot. During standard arraying operations, a large number of microscope slides are placed on and secured to a platter. DNA samples in 3x SSC in 96 or 384 well microtiter plates are placed on a stand. The robot positions a cluster of specialised spring-loaded printing tips into adjacent wells of the DNA source plates filling the reservoir slot of each tip with approximately 1 µl of DNA solution. The tips are then lightly tapped at identical positions on each slide leaving a small (<0.5 nl) drop of the DNA solution on the poly-lysine coated slide.

After depositing DNA on every slide the tips are washed and dried and the process is repeated for the next set of DNA samples with the new spots offset a small distance relative to previous spots to produce a high-density grid. The spacing between spots is determined by the size of the deposited DNA droplet, which is a function of the sharpness and characteristics of the tip and the hydrophobicity of the poly-lysine surface. Routinely, print arrays with spot-centre to spot-centre spacing of approximately 200 um and have achieved spacing approaching 100 um.

At 200 um spacing the entire yeast genome can be printed in a 1.8 cm square area, and at 100 um spacing all of the approximately 75,000 human genes can be printed on a standard 1" by 3" microscope slide. With the robot, each cycle takes about two minutes, and use a print-tip cluster with four tips spaced to fit in adjacent wells of standard 96 well microtiter plates. This results in a printing rate of 120 spots on 110 slides per hour. The cycle time can easily be cut in half with simple improvements in robotics, and we have a similarly sized print-tip cluster of 16 tips spaced for 384 well plates. This configuration results in a printing rate of approximately 1000 spots per 110 slides per hour, or six hours for the entire yeast genome. Expanding to 32 tips will enable a set of all human genes to be spotted on each of 110 slides in about two days.

As the only part of the robot that touches the DNA and the slides, the printing tips are the most critical component of making DNA microarrays. The tips we use operate on the same principle as a quill pen; liquid is drawn up by capillary action and deposited when the tip makes contact with the slide surface. The tips are made by first cutting a thin slot in a cylinder of stainless steel. One end of the cylinder is sharpened to a cone with the slot passing through the point. The two sides of the slot are squeezed together until they are almost touching at the point, and the tips are further sharpened by hand.

When the tips are immersed in the DNA solution, liquid is drawn up into the reservoir. When the tips are tapped gently onto glass, DNA exits only at the point. While many other factors can influence the size of the spot produced during printing, in general sharper tips produce smaller spots, so the tips must be kept as sharp as possible. Since we make all of our printing tips in a semimanual manner, they do not behave identically. Furthermore, while good tips can survive hundreds of thousands of taps without noticeable loss of quality, they do eventually need maintenance.

Thus, the first important step in printing good microarrays is to maintain high quality tips. Clean the tips by placing them in a bath sonicator with the tips floating, points submerged in clean water. Prime the tips by submerging the point in 3x SSC until it draws up liquid into the reservoir slot. Choose an initial set and place in the print head. Test the tips by doing a "test print". A test print is a print run where a single slide is placed on the platter, the tips are submerged in a fresh test solution containing 100 ng/μl salmon sperm DNA in 3x SSC and are tapped a few hundred times on the single slide with the spot-centre to spot-centre spacing desired for the ultimate printing.

After the test print, and examine the slide. The high salt content of the spotted DNA solution allows for easy visualisation of the size and shape of each spot. Carefully examine the slide to ensure that: 1) each tip is depositing DNA on the slide for at least the same number of taps as there are slides to be printed, 2) that the spots are of the desired size and shape and stay roughly the same size and shape across the slide, and 3) that all the tips are producing spots of roughly the same size and shape. When a tip fails to print, the cause is either a failure of the tip to load DNA, or a failure to discharge DNA onto the slide.

First, make sure that the tip is touching the glass when printing should occur. Second, make sure that the tip is loading DNA. This can be determined by removing the tip from the holder and slowly submerging it in the salmon sperm test solution described above while holding it up to the light. It should be possible to look through the slot to see if solution is being drawn into the slot. If the tip is not loading, make sure that the DNA solution is not too viscous, and then clean out the slot either by continued sonication or using a thin (<100 μm) stainless steel shim. If the tip is loading properly and is still not printing, it is likely that the tip is clogged. Remove the tip and examine it under a dissecting scope. The tip can be cleaned again either by sonication or by gently pulling a thin piece of metal through the end. Take care not to damage the tip's end. Make sure that the two ends of the tip are not touching, as this will prevent deposition of DNA.

If the spots are too large so that adjacent spots run together, the tips may not be sharp enough or they may be hitting the glass too hard. First try to reduce the impact of the tap, and then try sharpening the tips with fine sandpaper or a sharpening stone under a dissecting scope. Large spots can also be caused by insufficiently hydrophobic glass or by surfactant contamination in the DNA solution. While it is aesthetically pleasing for the spots produced by each tip to be the same size, variation in the size and shape of spots is not a serious functional problem so long as the spots remain sufficiently large, and well-separated.

If everything looks good in this test print, the process should be repeated at least once, and then the printing should be initiated immediately. Once begun, the printing process involves only feeding plates to the robot and vigilantly monitoring the quality of the printing. As the printing tips behave best when in constant use, printing should be interrupted as infrequently as possible. As the plates are open while they are being printed and will evaporate rapidly (changing the viscosity of the solution and potentially adversely impacting spotting), it is not advisable to design a platter to feed multiple plates to the robot unless it is accompanied by a mechanism either to remove the plate cover immediately before use, or to cool the plates to minimise evaporation. It is important that the printing process be closely monitored to make sure that spots are still being printed and that they are still an appropriate size and shape.

The most common problem is that a tip becomes clogged and stops printing. If this occurs, the print run should be interrupted immediately, and the problematic tip should be cleaned as described above. The test printing should be repeated until the tips are working well again. Another problem that occurs is that a dust particle or fiber gets stuck in the tip causing the spots to be irregularly shaped. This too should be cleared up by cleaning the tip.

A different type of problem occurs if the tips become dulled during the print, and the spots will become larger, potentially causing adjacent spots to overlap. This can be solved either by replacing the tip or by sharpening it. After printing is completed, the area containing spots should be delineated in some manner, as the spots will not be visible once the salt has been washed away. We generally do this by marking the boundaries of the area with a diamond tipped etching device. This pen should also be used to label the slides, as most other forms of labelling will be removed by the solvents used in post-processing. After the slides are labelled they should be removed from the platter and stored in a clean slide box in a cool, dry place. The slides are stable at this stage for at least a month, and need not be post-processed immediately.

Four steps are required to prepare printed arrays for hybridisation: rehydration and snap-drying, UV cross-linking, blocking, and denaturation.

Rehydration: The spotting process does not, in general, leave DNA evenly distributed throughout the spot. To more evenly distribute the DNA, the spots are rehydrated and snap-dried. This is a delicate process, as insufficient rehydration can produce irregular spot sizes and can reduce overall hybridisation intensity, while excessive rehydration can cause spots to fuse. Invert the metal block from a standard benchtop heating unit to produce a flat metal surface, and set the block to approximately 80C. Fill the reservoir of a plastic slide hydration chambers with hot tap water (approximately 50-60C).

Process each slide separately by placing it, array side down, over the reservoir, allowing the vapour to hydrate the elements until they all are glistening. Immediately place the slide, array side up, on the heat-block until the spots have dried. This is a process that requires practice. More consistency can be achieved by carrying out the hydration over a temperature controlled water bath, although even then the amount of time required to hydrate different arrays, even within one batch, can vary considerably. Some groups prefer to hydrate at lower temperatures for longer periods of time.

UV Crosslinking: Once the arrays have been rehydrated and snapdried, the DNA is crosslinked to the slide by UV irradiation. While not essential, this step can increase the amount of hybridisable DNA stably attached at each spot, especially when the DNA concentration in the spotting solution is low. UV crosslinking is carried out by placing the slides, array side up, on plastic-wrap, in a Strategene UV 1800 Stratalinker. Switch from timed mode to total energy mode and expose the slides to 60 mJ of energy.

Blocking: The most critical step in post-processing is the blocking step, where the remaining free lysine groups are modified to minimise their ability to bind labelled probe DNA. If these groups are not blocked, labelled probe will bind indiscriminately and non-specifically to the surface and will produce excessively high background. We block by acylation with succinic anhydride. The charged amino group of the lysine carries out a nucleophilic attack on one of the carboxyl carbons, forming an amide bond and exposing a free carboxylate group on the other end of the chain.

In addition to converting the amino groups of the lysines to amides, this new moiety creates a negatively charged surface which further reduces non-specific binding of DNA. As the blocking reagent has limited stability in solution, it is important that this process be carried out quickly. Find a glass chamber, and a slide rack with a handle that will fit easily in this chamber. Place slides in this rack, and place the rack in a fume hood right next to the chamber.

The chambers we use hold 350 ml of solution, scale the following reagents to the volume you will need to completely cover the slides. Measure out 335 ml of water and pour into a 500 ml glass beaker. Mark the level of the solution with a pen on the outside of the beaker. Remove the water and rinse the beaker with 95% ethanol. Carefully dry the beaker with a paper towel, place a small magnetic stir bar in the beaker, and place on a stir-plate in a fume hood. Measure 15 ml 1M sodium borate pH 8.0 and place next to the beaker. Weigh out 6g of succinic anhydride and place in the beaker.

Quickly fill the beaker to the 335 ml mark with 1,2-methyl pyrrolidinone and begin stirring at high speed. Immediately after of the succinic anhydride is dissolved, add the sodium borate and stir until the solutions mix (about 5 seconds). Pour the solution into the empty slide chamber, immediately dunk the slides into the solution and plunge the slides up and down 5 times. Cover the rack and shake gently for 15 minutes.

Denaturation: While (or before) the blocking reaction is taking place, prepare a large beaker or glass chamber filled with boiling ddH_2O. The volume should be at least twice that of the blocking solution. Turn off the heat immediately shortly before the blocking reaction is complete, and when the water is no longer bubbling, quickly transfer the slides to the water, plunge the slides up and down 3 to 5 times and allow the slides to sit for 2 minutes. Immediately transfer the slides to a chamber containing 95% ethanol, and again plunge the rack up and down 3 to 5 times. Transfer the slides to a centrifuge and spin dry. The slides are now ready for hybridisation.

Probe Preparation

Isolation of RNA: We have had success with RNA prepared by a variety of methods including hot phenol extraction, guanidinium isothiocyanate stabilisation and others. We have had good success using Invitrogen's FastTrak 2.0 mRNA isolation kits for both yeast and human samples. The main criterion is the production of high quality RNA. For gene expression studies, the best results have been achieved using purified polyadenylated RNA as the template for reverse transcription, although, at least with yeast, total RNA has been used with good results. A good experiment requires at least 1μg of poly-A RNA (or 100 μg total RNA) per sample as template for synthesis of fluorescent cDNA probe. Methods for amplifying smaller amounts of RNA are being examined for their compatibility with the methods described below.

Labelling: A variety of methods have been used for preparation of labelled probe from RNA. In most cases we label cDNA directly by incorporating fluorescently labelled nucleotides during oligo-dT primed reverse transcription. Where appropriate, random-primed cDNA synthesis can be used for labelling, although it may not be appropriate for all array designs. For example, arrays composed of sequences from the 3′ ends of transcripts are more suited for analysis of oligo-dT primed cDNA probes, while arrays whose elements consist of entire cDNA sequences or predicted open reading frames are well suited for probes prepared by either priming method. Samples can also be labelled during second-strand synthesis after first-strand synthesis of unlabelled cDNA. Protocols for all three methods are given below.

Fluorescent dyes: A large number of fluorescently labelled deoxyribonucleotides are now commercially available. Excellent results have been obtained with Cy3- and Cy5-dUTP; they are spectrally well separated, can be incorporated with high specific activities with a variety of enzymes, and brightly fluoresce when dry which simplifies the image acquisition process.

Oligo-dT and random primed labelling: All reagents should be sterilised and DEPC treated. Start with RNA in 10 µl H_2O. This solution should include any control RNAs. Add 2.5 µl (6 µg) primer. Incubate in a 70C water bath for 10 minutes to denature RNA. Place on ice. Mix 2 µl MMLV reverse transcriptase (from GibcoBRL SuperScriptII Rnase H- Reverse Transcriptase kit; catalogue number 18064-014), 6 µl 5x first-strand buffer, 3 µl 0.1 M DTT, 6 µl 5x nucleotides (2.5 mM dATP, 2.5 mM dCTP, 2.5 mM dGTP, 1.0 mM dTTP), 3 µl 1mM Cy3 or Cy5 linked dUTPs (Amersham). Add to RNA-primer solution to a final volume of 30µl.

Incubate at 42C for one hour. Add 1 μl SuperScriptII reverse transcriptase and incubate at 42C for an additional hour. Add 1.5μl 20 mM EDTA to stop the reaction and 15 μl 0.1 M NaOH to degrade the template RNA. Incubate at 70C for 10 minutes. Immediately add 15μl 0.1 M HCl to neutralise the sample (the fluors, Cy5 in particular, are sensitive to high pH). Place the samples in a Microcon 30 microconcentrator (Amicon) and add 500 μl TE (10 mM Tris, pH 8.0, 1 mM EDTA). Spin 5 to 10 minutes in a benchtop microcentrifuge at high speed to a volume of 10 to 20 μl.

Discard the flow-through. Add 500 μl TE and repeat the spin. After the second spin, the probe retained by the Microcon should be significantly brighter than the flow-through. This is a very strong indicator of a successful labelling reaction. If no colour is retained, it is very unlikely that the labelling will produce a good hybridisation result. Collect the probes by inverting the filter and spinning for 1 minute. Mix the appropriate Cy3 and Cy5 labelled probes and bring up in 500 μl TE. Spin to final volume of 8 μl and collect probe (which should now be purple). Add 2μl of a blocking solution containing 10μg/μl oligo-dA 20-mers, 10μg /μl yeast tRNA and, for human probes, 10μg/μl human cot-1 DNA.

Add 2.1 μl 20x SSC (to a final concentration of 3.5x) and 0.4 μl 10% SDS (to 0.3%). Be careful that no SDS is present on the outside of the pipette tip when SDS is added to the sample, as extra SDS can adversely effect the hybridisation. Denature the sample by placing in a 100C water bath for 1 minute. Allow the sample to cool on the benchtop. For human samples the sample should be left at room temperature for 30 minutes to allow for cot-1 hybridisation to repetitive elements. For yeast, the probe is ready for hybridisation as soon as it cools. 12 μl is an appropriate volume for arrays around 2cm × 2cm. A larger final volume should be used for arrays covering a larger area.

Klenow labelling of cDNA: Prepare a cDNA template using the above procedure; except use a 5x nucleotide solution containing 2.5 mM of each dNTP (i.e. final concentration of 0.5 mM each dNTP in the labelling reaction) and do not include fluorescent nucleotides. Clean up the sample as above to a final volume of 20 µl. Add 3 µl of a random hexamer or nonamer (4 µg/µl). Heat to 100C for one minute. Let cool on benchtop for 5 minutes. Add 5 µl 10x buffer (100mM Tris-HCl, pH 7.4, 50 mM $MgCl_2$, 75 mM DTT), 5 µl 10x nucleotides (0.25 mM dATP, 0.25 mM dCTP, 0.25 mM cGTP, 0.09 mM dTTP), 1.5 µl 1mM Cy3 or Cy5 dUTP (Amersham), 2 µl (10 activity units) exonuclease free klenow, and 13.5 µl ddH_2O to a final volume of 50 µl. Incubate at 37C for 4 hours. Clean up and prepare for hybridisation as above.

Hybridisation

Spin the probe sample for one minute in a benchtop microcentrifuge at high speed to pellet any particulate matter. Pipette the probe onto the centre of the array, being careful not to dislodge any pelleted material in the tube and to avoid forming bubbles. Place a cover slip (large enough to cover the entire array surface) over the probe, again being careful to avoid the formation of bubbles. Pipette three 5µl drops of 3x SSC on a separate part of the slide to provide humidity in the hybridisation chamber and thus ensure that the probe mixture does not dehydrate during hybridisation). Place the slide in a sealed chamber and submerge in a 65C water bath for 4 to 16 hours.

Following hybridisation, carefully dry the outside of the slide chamber and remove the slide. Place the slide in a slide rack submerged in the wash solution 1 with the array face of the slide tilted down so that, when the cover slip drops off, it does not scratch the array surface. After the cover slip comes off, plunge the slide up and down in this solution 2-3 times and transfer to a new slide rack submerged in wash solution 2. Be careful to minimise the transfer of wash solution 1 to the second chamber as SDS can interfere with slide imaging. Gently rock the chamber for 2 minutes, transfer the slide to wash solution 3 for 2 more minutes, then spin dry the slide. The wash solutions should all be at room temperature. Altering the temperature and/or time of these steps will change the stringency of the hybridisation.

Because of the highly parallel nature of microarray experiments, it is possible, and highly valuable, to include a large number of controls on every array and in every hybridisation. Some useful controls are:

Doped RNAs: It is very useful to dope in known amounts of control RNAs at various stages of the probe preparation process to monitor and assess various aspects of each experiment. Clearly, doped RNAs should be chosen to have no cross hybridisation with RNA from the organism being studied, but should be similar in their general characteristics. Many yeast genes can be used as controls on human gene expression arrays, as there is limited homology at the nucleotide level between yeast and humans. Hundreds of such genes can be spotted on human arrays with only a nominal increase in the time required for arraying.

By cloning these genes into bacterial plasmids with phage RNA polymerase binding sites and artificial poly-A tails, large amounts of RNA can be made for each gene for incorporation at various experimental stages. Such controls can be used in a number ways. For example, RNA from a number of such controls can be doped in at various ratios between the Cy3 and Cy5 samples and at varying total concentrations to assay the stringency of the hybridisation, the calibration of output ratios to input ratios and the sensitivity of detection.

RNA quality controls: The amount of a given nucleic acid species that will bind to a complementary target element can be sensitive to changes in the size of the complementary region. Thus, the intensity of the hybridisation signal for a given cDNA can vary with differences in RNA quality. This can confound microarray hybridisation results as, for example, an apparent non-unity ratio can be observed when equal amounts of a given cDNA derived from RNA of differing quality are compared. Judicious choice of target elements can minimise this problem; for oligo-(dT) primed cDNA probe, target elements restricted to the 3' ends of genes will be minimally sensitive to RNA quality.

RNA quality can be easily monitored with appropriate control elements. For example, target elements of DNA fragments of 100-200 basepairs tiled across an entire gene can be used to assay the relative RNA quality of that gene in the samples being compared degraded RNA, labelled by oligo-dT primed cDNA synthesis, will give a proportionally weaker signal for the more 5' target elements. The experimental step that is the source of problems with RNA quality may be identified by making such tilings of genes used as doped controls, and then adding the corresponding RNA at various stages of the RNA and probe preparation process. For example, RNAs can be doped in when cells are harvested, when poly-A RNA is isolated, when labelled probe is made, and immediately before hybridisation.

Blocking controls: Most hybridisation reactions require the addition of cold DNA to block hybridisation due to non gene-specific sequence similarities between probe and target, especially those due to repetitive elements. Control target elements corresponding to such potentially confounding sequences can be used to monitor the success of the blocking. For example, on human gene expression arrays, a human cot-1 DNA spot should have little or no hybridisation signal if repetitive elements have been successfully blocked. Other useful negative controls include oligo-dA, oligo-dT, plasmid DNA and human genomic DNA.

Normalisation controls: As it is difficult to equalise accurately the amount of RNA used to prepare labelled probe for multiple samples, it is useful to have control target elements whose hybridisation signal is expected to be the same for both samples used in a hybridisation. For example, for yeast gene expression experiments, target elements of total yeast genomic DNA can be used to normalise the signal between the two samples. Human gene expression experiments are more complicated in this regard; as only a small fraction of the human genome is expressed, the hybridisation signal to total human genomic DNA should be very low.

After hybridisation, a fluorescent image of the array must be acquired for both fluorescent dyes used. The device we use is a laser scanning confocal microscope (scanner). The scanner has a laser (or lasers) producing light with a wavelength appropriate for the excitation spectra of the dyes being used (for Cy3 this is green light around 540 nm; for Cy5 red light around 650 nm). The light passes through a standard microscope objective and illuminates a single point on the slide.

The emitted light gathered by the objective passes through a series of filters (to remove the excitation beam), a collimating lens, and a pinhole and is quantified in a photomultiplier tube. The slide is rapidly scanned over the laser beam and a raster image of the array is acquired. It is possible to acquire a similar image using a fixed slide and a pair of galvanometers to position the beam at all possible points on the array surface, or using a wide-field illuminating source and a CCD camera. The confocal scanning device gives the best results as the signal to noise ratios are superior to those achieved with galvanometer or CCD based devices.

The experimental goal is to measure, for each spot on the array, the relative amount of fluorescence from each fluor used. The are many ways to accomplish this, the most obvious being segmenting the image into areas corresponding to each spot and them determining the average fluorescence intensity in each box for each fluor. The relative amount of each element in the two samples is just the ratio of the average fluorescent intensities. This kind of calculation is confounded somewhat by issues involving the compensation for the nonzero fluorescent background observed on most arrays.

Technical Problems

Experiments using DNA microarrays, can encounter technical problems at any step. The opportunity, provided by the microarray format, to include numerous internal controls, can greatly facilitate the recognition and correction of many kinds of problems. Indeed, even severely flawed experiments often provide plenty of useful new information. Still, it is wise, whenever possible, in designing an experiment, to plan so that any step of the experiment can be repeated without having to repeat the experiment in its entirety. For example, it is always wise to prepare more mRNA and more arrays than would be necessary if all went well. Few microarray images are perfect.

Many problems that we have encountered are difficult to trace, and do not manifest themselves in detectable ways until the array is scanned. It is advisable to perform multiple levels of quality control before hybridising valuable experimental samples. Batches of poly-lysine coated glass should be tested by printing small arrays that are taken through the entire post-processing and hybridisation process. Only batches of slides that have been used successfully for such small arrays should be used for "real" array printings.

Once an array has been printed, post-processing should be performed on small batches of arrays and the success of the postprocessing should be determined by performing a test hybridisation with probe prepared from RNA that is known to produce good results. Finally, all arrays should be scanned prior to hybridisation to eliminate slides with obvious blemishes and to eliminate arrays with improperly high fluorescence prior to hybridisation. Poly-lysine and some other coatings appear to fluoresce weakly at wavelengths used to detect Cy3; for unknown reasons this fluorescence varies from array to array and can occasionally reach intensities that will interfere with the hybridisation signal.

In addition, printed DNA also fluoresces weakly at these wavelengths and this fluorescence also varies from array to array. If either of these pre-hybridisation sources of fluorescence are at levels near those of the signal on hybridised arrays, these arrays should be discarded. Some of the more common problems are described below.

Spot shape problems: One common problem is that some or all of the spots on the array have "comet tails", where the fluorescence of a spot tails off in a direction that is consistent across the array and where the colour of a spot is reflected in the material tailing away from the spot. This phenomenon is likely caused by insufficiently rapid immersion of the slides in the succinic anhydride blocking solution, resulting in DNA binding to the slide as it is being washed off the spot.

To the extent that spots can be resolved from each other arrays with comet tails are still usable. Another problem related to general spot appearance are "donut holes". When DNA is deposited using tips that those described above, it is not left in a uniform circle; far more material accumulates near the outer edge of the spot which leaves a relatively empty area in the centre. This phenomenon can usually be eliminated by longer rehydration. Even more so that comet-tails, donut holes do not make an array unusable and in fact only slightly compromise data quality.

High Background: There is a nebulous class of array problems marked by abnormally high fluorescent backgrounds. High backgrounds are observed across the entire hybridised surface as well as in localised regions on the array. Many potential sources of high background can be described, but it is difficult to diagnose the source of a particular type of high background. In attempting to diagnose background problems, it is important to determine if the background is restricted to the area of the slide that was exposed to probe.

Unusually high background that covers the whole slide can usually be avoided if arrays are systematically scanned prior to hybridisation. High background restricted to the hybridised area indicates that labelled probe is adhering to the slide surface such that it is not being removed during the washing steps. Two possible causes of such an effect are insufficient blocking of the amino groups of the poly-lysine coating, or precipitation of the labelled probe during hybridisation. High background from these sources is usually accompanied by a phenomenon known as "anti-probe", where there the high fluorescent background occurs only in places where no DNA was spotted. This likely reflects the diminished capacity of the poly-lysine groups in areas containing spotted DNA to bind additional DNA.

In some cases, the background is clearly worse towards one side of the array, indicating a relationship to how the slide was positioned in a liquid medium. Often, high background is observed near the edges of the coverslip, likely indicating that the probe dehydrated during the hybridisation. This can be corrected by adding a larger volume of 3x SSC to the hybridisation chamber and ensuring that the chambers are well sealed. It is also possible for uniformly high backgrounds to result from poor probe preparation, as small probe fragment and/or unincorporated fluorescent nucleotides can bind to the surface and contribute to background without contributing to the hybridisation signal.

Locally weak signal: Most of the phenomena that produce high background are also accompanied by a diminution of specific signal, as much of the activity of the probe is being sequestered by the array surface. However, locally decreased signal that is not accompanied by high background is also observed. This is due to a local decrease in the thickness of the probe layer in that region. Air bubbles and inhomogeneities in the flatness of the slide and/or cover slip can cause such an effect. Care in choosing and positioning the cover slip can minimise this problem.

GENE EXPRESSION PROFILING

Expressed sequence tags (ESTs), microarrays, large-scale gene expression profiling, and associated informatics technologies are rapidly becoming commonplace in the plant sciences. These 'genomic' approaches typically take advantage of technologies for characterising large numbers of nucleic acid sequences, bioinformatics, and the expanding collection of nucleic acid sequence data from diverse taxa. 'Systems biology' attempts to combine large-scale DNA sequence, gene expression, protein, metabolite, genotype, and/or phenotype data to develop a comprehensive understanding of biological process. Combination of these approaches also makes it possible to extract more meaningful functional information as new DNA sequence data are generated. While the potential of this technology is enormous, the utility of large-scale expression data is not always well understood, nor are the limitations in analysis and interpretation.

Comprehensive transcriptome analysis should make it possible to identify and dissect complex genetic networks that underlie processes critical to physiology, development, and response. For example, gene regulatory networks have been inferred using microarray data obtained from a variety of organisms, and it is also possible to correlate these genetic networks with metabolic processes. Gene networks are also being dissected in plants for processes ranging from seed filling to cold tolerance.

Further characterisation of gene networks in plants will help us to understand the molecular basis of plant processes and identify new targets for manipulating biochemical, physiological, and developmental processes in crop species. Nevertheless, a comprehensive characterisation of the transcriptome is not a prerequisite for studying every biological question and the value of such an approach can only be weighed in light of a clearer understanding of the possibilities and caveats of transcriptome profiling.

Rob Alba et. al, done a study for gene expression profiling in plants with emphasis on cDNA microarrays. Datails of their study is briefly described below. Studies such as these are generating an overwhelming amount of microarray data, the vast majority of which has yet to be analysed sufficiently or integrated with our existing knowledge base.

Expressed Sequence Tags (ESTs)

Expressed sequence tags are created by sequencing the 5′ and/or 3′ ends of randomly isolated gene transcripts that have been converted into cDNA. Despite the fact that a typical EST represents only a portion of a coding sequence, en masse this partial sequence data is of substantial utility. For example, EST collections are a relatively quick and inexpensive route for discovering new genes, confirm coding regions in genomic sequence, create opportunities to elucidate phylogenetic relationships, facilitate the construction of genome maps, can sometimes be interpreted directly for transcriptome activity, and provide the basis for development of expression arrays also known as DNA chips.

In addition, high-throughput technology and EST sequencing projects can result in identification of significant portions of an organism's gene content and thus can serve as a foundation for initiating genome sequencing projects. Currently there are nearly 20 million ESTs in the NCBI public collection, more than 4 million of which derive from plants. With many large-scale EST sequencing projects in progress and new projects being initiated, the number of ESTs in the public domain will continue to increase in the coming years. The sheer volume of this sequence data has and will continue to require new computer-based tools for systematic collection, organisation, storage, access, analysis, and visualisation of this data.

Not surprisingly, despite the relative youth of this field, an impressive diversity of bioinformatics resources exists for these purposes. As sequence and annotation data continue to accumulate, public databases for genomic analysis will become increasingly valuable to the plant science community. The Arabidopsis Information Resource, the Salk Institute Genomic Analysis Laboratory, the Solanaceae Genomics Network, and GRAMENE serve well as examples of these on-line resources. In addition to a variety of analysis tools and a wealth of microarray data, TAIR contains sequence data for the entire Arabidopsis genome that is easily accessible via query and FTP tools. All 120 Mb of this sequence data can also be obtained from the SIGnAL database, as can sequence data for more than 11 000 full-length cDNA sequences, and more than 10 000 publicly available ORF clones. SIGnAL also curates a mapped collection of more than 300 000 sequenced indexed T-DNA insertion mutants, most of which are publicly available through the Arabidopsis Biological Resource Center.

SGN is dedicated to the biology of Solanaceae species, including tomato, potato, tobacco, eggplant, pepper, and petunia. This database contains curated sequence data derived from nearly 300 000 ESTs, extensive mapping data for the tomato genome, in addition to mapping data for the genomes of potato and eggplant. GRAMENE is a curated, open-source database dedicated to the biology of grasses, including rice, maize, Sorghum, barley, and wheat. The primary objective of GRAMENE is to facilitate cross-species homology relationships among monocotyledonous species. In addition to sequence and mapping data, GRAMENE also contains a variety of analysis tools, an extensive Ontology database, a QTL database, and a database for mutant genes in rice.

With databases such as these, and advances in computational molecular biology and biostatistics, it is possible to mine and analyse large EST datasets efficiently and exhaustively. Particularly important is the fact that this type of data mining can be used to corroborate and extend upon expression data obtained from microarray experiments. Using such a strategy Rob Alba et. al analysed a tomato EST dataset representing 152 635 ESTS to gain insights into statistically significant changes in differential expression among diverse plant tissues representing a range of developmental programmes and biological responses.

By clustering genes according to their relative abundance in the various EST libraries, expression patterns of genes across various tissues were generated and genes with similar patterns were grouped. In addition, tissues themselves were clustered for relatedness based on relative gene expression as a means of validating the integrity of the EST data as representative of relative gene expression. EST collections from other species were also characterised to facilitate cross-species comparisons where possible.

With the rapid expansion of available EST data, opportunities for digital analysis of gene expression will continue to expand. Expressed sequence tag collections also have limitations when being used for genomic analysis from the perspectives of accurate representation of genome content, gene sequence, and as windows into transcriptome activity. The fact that ESTs reflect actively transcribed genes makes it difficult to use EST sequencing alone as a means of capturing the majority of an organism's gene content. Additionally, and of great importance, is the fact that a fraction of this sequence data is erroneous. Some of these sequence errors derive from the imperfect nature of the enzymes used to generate cDNA libraries and sequence data.

At present these sequence errors cannot be completely avoided, but multiple sequence reads through the same gene makes it possible to minimise this type of artifact. EST collections, even when not normalised or subtracted, are not perfectly representative of the mRNA populations they originate from. For example, low-abundance transcripts are unlikely to be represented fully in all EST collections. Misrepresentation can also originate from transcripts with atypical sequence features that impair reverse transcription and/or subsequent cDNA cloning. A third source of error in EST collections can originate during processing of sequence data. This type of sequence error derives from the imperfect technology and algorithms used for base calling, sequence annotation, and contig assembly.

Finally, EST sequence data are also prone to human errors during storage, handling, replication, and management of EST collections. Consequently, re-sequencing ESTs of interest is an important means of validation prior to further characterisation. Despite these limitations, it has been shown that EST databases can be a valid and reliable source of gene expression data.

Gene Expression Profiling

A variety of methods have been developed for quantifying mRNA abundance in plant tissues. Although the established and reliable method of RNA gel-blot analysis can be quite sensitive and allows for the accurate quantification of specific transcripts, this method is not readily adapted to genome-scale analysis. Differential display uses lowstringency PCR, a combinatorial primer set, and gel electrophoresis to amplify and visualise larger populations of cDNAs representing mRNA populations of interest. Differential display has important advantages when compared with scale-limited approaches such as RNA-blot analysis, yet this technique suffers from output that is not quantitative and positives are often difficult to clone and confirm.

More recently the principles of AFLP® have been applied to cDNA templates and this approach has been used to identify differentially expressed genes involved in a variety of plant processes. This technique offers several advantages over more traditional approaches. Of particular importance is the fact that poorly characterised genomes can be investigated in a high-throughput manner. Because the stringency of cDNA-AFLP PCR reactions is quite high the fidelity of the cDNA-AFLP system allows much greater confidence in acquired data and differences in the intensities of amplified products can be informative.

In addition, this technique allows a wide variety of tissue types, developmental stages, or time points to be compared concurrently. As with the other profiling methods described here, the sensitivity of cDNA-AFLP is only limited by the ability of cDNA libraries to capture low-abundance transcripts. Sequencing of cDNA libraries is a more direct and comprehensive approach to gene expression profiling, but this method requires substantial resources for cloning and sequencing, and is less sensitive to low-abundance transcripts. Serial analysis of gene expression is an elegant technique that combines differential display and cDNA sequencing approaches, and it has the advantage of being quantitative.

Unfortunately, SAGE is laborious, requires an extensive foundation of sequence information, and suffers from some of the same concerns regarding low-abundance transcripts. Microarrays take advantage of existing EST collections and genome sequence data, robotic instrumentation for miniaturisation, and fluorescent dyes for simultaneously detecting nucleic acid abundance in RNA populations derived from multiple samples. Populations of fluorescent cDNA targets representing the mRNA populations of interest are queried via hybridisation with a large number of probes that have been immobilised on a suitable substrate.

The arrays themselves are composed of collections of DNA sequences that have been printed as a microscopic grid of catalogued features by a high-fidelity robotic system. This technique for gene expression profiling has important advantages when compared with RNA-blot analysis, cDNA sequencing, differential display, AFLP analysis, and SAGE. Most importantly, it can measure tens-of-thousands of different mRNA transcripts in parallel, it is semi-quantitative, and it is sensitive to low-abundance transcripts that are represented on a given array. This last point is worth emphasising in that microarrays are inherently limited to their contained sequences, while so-called 'open architecture' systems such as differential display and SAGE can capture information for any sequence that is expressed at a level sufficiently above the level of detection.

In those instances where complete genome sequence is available, microarrays make it possible to monitor the expression of an entire genome in a single experiment. Despite this potential, predominant uses of microarrays facilitate analysis of significant, yet limited, subsets of the target genome. When used for time-course analyses, analyses of transcriptome alterations caused by genetic lesions, or comparison of transcript accumulation in similar tissues from closely related species, the potential of microarrays for gene expression profiling is not only enormous, but is also just beginning to be realised.

Microarrays

Two different types of microarrays have become commonplace: cDNA microarrays and oligonucleotide microarrays. Both have notable and distinct advantages. For example, cDNA arrays can be prepared directly from existing cDNA libraries, a large number of which are in the public domain. Thus, fabrication of cDNA arrays is only dependent upon availability of ordered clone collections, and appropriate arraying and scanning instrumentation. Once a set of corresponding PCR products has been generated, arrays can be created in multiple versions containing the entire set of available sequences or subsets of sequences resulting in smaller 'boutique' arrays suitable for specific research applications.

Smaller 'boutique' arrays are also useful for reducing a statistical problem of scale. Here, a large array might be used to identify differentially expressed genes of interest, which could then be re-arrayed as a smaller array and used in subsequent experiments. One benefit of this approach is that it can free up resources that can be used to increase experimental replication and thereby increase precision. Another advantage of cDNA arrays is that they can be used in 'two-colour' co-hybridisation experiments that allow direct comparisons of transcript abundance in two mRNA populations of interest.

Although this strategy generates comparative expression ratios instead of measuring absolute expression levels, it is effective for comparative expression profiling and reduces experimental variation that arises in microarray data collected from different chips. Like cDNA arrays, oligonucleotides can be printed using robotic instrumentation and sub-arrays for specific research applications can be fabricated easily. The main limitations in development of oligonucleotide arrays are the costs associated with sequence selection and oligonucleotide synthesis. As these costs continue to decline oligo-based arrays are likely to become more predominant in the near future because they offer a number of important advantages over cDNA arrays.

One such advantage is the fact that oligobased arrays can be fabricated using microfluidic technology, which utilises light to direct the synthesis of short oligonucleotides onto a suitable matrix. Photolithography is particularly useful because it allows for the fabrication of extremely high-density arrays. Another important advantage is that the probes in an oligonucleotide array are designed to represent unique gene sequences such that cross-hybridisation between related gene sequences is minimised to a degree dependent upon the completeness of available sequence information.

Cross-hybridisation between homologous sequences continues to be problematic when using cDNA arrays. Furthermore, the array elements in an oligonucleotide array are typically designed to have uniform length, uniform melting temperatures, and to be of uniform concentrations, which can significantly reduce experimental variation and thereby increase statistical power and precision. The primary disadvantage of oligo-based arrays is that oligonucleotide sets can be very expensive because of the extensive sequence data and computational input required for designing gene-specific oligonucleotide probes.

Currently, a single oligonucleotide chip is often three to five times more expensive than the cost of a single cDNA chip. A common limitation of all array approaches is the requirement of significant RNA for the preparation of fluorescently labelled targets. For this and other reasons, methods for generating sufficient signal from extremely small RNA populations have and will continue to be investigated.

More recently, experimental designs for microarray analyses have begun to incorporate interspecies comparisons using arrays that originate from one of the genomes being investigated. Toward this objective, comparison of closely related species is most informative because artifacts stemming from sequence divergence are minimised. As an example of this type of interspecies comparison the researchers co-hybridised labelled cDNA populations derived from pepper pericarp and tomato pericarp to the TOM1 microarray. Pepper genes showing increased transcript abundance in this experiment (compared with expression in equivalent tomato tissue) are of particular interest because this result cannot be explained by differential hybridisation because of sequence divergence.

Verification of microarray data can be accomplished in a variety of ways, including RNA-blot analysis, RT PCR, realtime PCR, and/or comparison with EST expression databases. The latter being the only approach that has potential for genome-scale verification. The comparative profiling experiment described above identifies a number of genes that have been previously shown to be highly expressed in tomato pericarp during ripening, including genes encoding ACC oxidase1, pectin esterase, polygalacturonase 2A, pectate lyase, histidine decarboxylase, and acid invertase.

EST data in the Tomato Expression Database again confirms that all six of these genes are abundant in breaker-stage pericarp tissue. In addition, 26 genes were identified that (i) are highly expressed in the breaker-stage library of the Tomato Expression Database and (ii) exist on the TOM1 microarray. Twentyfour of these 26 genes (92%) sorted into the top 50% of fluorescent signal intensities when assayed using the TOM1 microarray. The continued expansion of

EST collections will only increase the feasibility of this approach for large-scale verification of transcriptome data. Differential expression of genes identified via comparative expression profiling can also be used in conjunction with large-scale proteome analysis to dissect regulatory processes.

Comparisons that combine expression and proteome profiling should allow one to distinguish transcriptional versus post-transcriptional regulation. To date, however, the potential for using comparative genomics/proteomics for this type of molecular analysis in plants remains largely untapped.

In most cases biological replication is superior to technical replication, and technical replication is far better than none. Consistent with the proposal of Lee et al., for timecourse experiments they suggest a minimum of three to four biological replicates for each time point with a dyeswap technical replicate for each biological replicate. This approach minimises dye-specific artifacts and makes statistical analyses possible. Dye-swap replicates, which involve repeating the hybridisation conditions with dye reversal in the second hybridisation, are useful for reducing systematic dye bias. Presumably this dye bias derives from differences in mean brightness and background noise, dye-specific incorporation efficiency, different extinction coefficients, differences in quantum fluorescence yield, and other physical properties of the dyes.

For experiments comparing only two or three different samples, they suggest a minimum of three to six biological replicates to make statistical analysis possible while minimising resource depletion. Dye-swap replicates should be used in this situation as well. It is also important to point out that they are not implying that the 'law of diminishing returns' has firmly set in after three to six replicates. In some cases, particularly scenarios where more precision is required or greater systemic variance has been documented, greater replication is likely to be worth the additional resource investment. Thus, three to six replicates should be considered only as an initial guideline during experimental design.

Although biological replication is most desirable, in some cases such replication is not possible. In these instances, consistent with the hypothesis of Peng et al., they continue to presume that a single RNA extraction from a homogenous pool of replicate tissue is superior to a single RNA extraction from an individual tissue sample. Labelling of cDNA targets and microarray hybridisation Purity and integrity of RNA can influence cDNA synthesis, incorporation of fluorescent dyes, dye stability, and probetarget hybridisation. In addition, impurities such as genomic DNA, cellular proteins, lipids, and carbohydrates can lead to non-specific binding of fluorescent cDNAs to array elements and chip surfaces. Thus, methods for RNA extraction are an important consideration for microarray studies and optimised protocols can differ notably for different types of tissues, experimental designs, and/or experimental questions. For example, some experimental questions require protocols optimised for high throughput, others require protocols optimised for very low-abundance transcripts, while others require protocols optimised for RNA yield.

A variety of techniques have been used successfully for purifying RNA from plant tissues prior to use in microarray experiments, including phenol-based extraction methods, guanidine thiocyanate, TRIzol®, silica-based RNA extraction, and methods that use proprietary extraction cocktails such as RNAwiz. Protocols developed specifically for quantitative extraction of low-abundance transcripts are also likely to work for many plant tissues. The work frequently employs a modified version of the protocol reported by Chang et al., which was designed for use with pine tissues that are rich in carbohydrates and secondary metabolites.

Modifications include purification of total RNA from 3.5 g of frozen tissue powder using 17 ml of extraction buffer, additional chloroform:IAA extraction steps prior to the LiCl precipitation, and elimination of the chloroform:IAA extraction steps after the LiCl precipitation. When mRNA is desired, the EtOH-washed total RNA is brought to 0.75 lg $\mu l)^{-1}$ with 10 mM Tris-Cl (pH 7.5) and the mRNA is purified with oligo $d(T)_{25}$ Dynabeads™, as per the manufacturer's protocol. They find that fluorescent cDNA targets prepared from mRNA usually result in reduced background fluorescence and stronger more consistent signal across the array.

Regardless of the extraction protocol used for RNA isolation, purity and integrity should always be assayed prior to cDNA synthesis and cDNA labelling. RNA purity can be assessed via absorbance measurements at 230, 260, and 280 nm. The 260:280 ratio is indicative of protein contamination and the 260:230 ratio is indicative of contamination by carbohydrates or other metabolites. In both cases, ratios below 1.9 can be problematic for cDNA synthesis/ labelling and ratios greater than 2.1 are optimal. RNA integrity can be assayed using formaldehyde denaturinggel electrophoresis or the Agilent 2100 Bioanalyzer system.

The two most common methods for labelling cDNA targets are direct incorporation of fluorescent nucleotides during reverse transcription, and a two-step incorporation method often referred to as indirect incorporation or aminoallyl labelling. Direct incorporation was initially the method of choice, but persistent problems with this method have been reported. For example, the rate of incorporation for nucleotides labelled with cyanine3 (Cy3™) and cyanine5 (Cy5™) is typically low, can be influenced by cDNA sequence, and can have negative effects on cDNA yield—all leading to inaccurate representations of expression. Indirect incorporation is the emerging method of choice for labelling cDNA prior to microarray hybridisation. This method utilises dUTP nucleotides that are modified with an aminoallyl group.

After incorporation of these nucleotides during cDNA synthesis, the modified cDNA is labelled using an *N*-hydroxy-succinimide ester form of Cy3™ or Cy5™ and a carbonate-based coupling buffer. This approach circumvents low incorporation rates and incorporation bias that are likely due to the size of the dye molecules. They had good results with a commercially available version of the amino-allyl labelling method for preparation of fluorescent cDNA targets. This labelling system uses two types of modified nucleotides, one of which contains an amino-allyl modification and one of which contains an amino-hexyl modification.

In theory, this approach should increase the number of available nucleotides that can be coupled to dye molecules, thereby increasing specific activity and enabling greater sensitivity. Some protocol modifications used for this cDNA labelling method are:

(i) overnight precipitation of the transcribed cDNAs,

(ii) an additional wash step during the cDNA purification,

(iii) incubation of the 2X coupling buffer at 37°C to ensure that reagent precipitates are completely dissolved,

(iv) incubation of the coupling buffer/ cDNA suspension at 50°C for 10 min followed by thorough mixing to ensure the cDNA pellet is completely re-suspended prior to the labelling reaction,

(v) two additional wash steps during purification of the labelled cDNA, and

(vi) elution of labeled cDNA with 63 ml of nuclease-free H_2O.

Success of the cDNA synthesis and dye labeling reactions should always be assessed spectrophotometrically.

Frequency of incorporation should also be calculated (frequency of incorporation = pmol cDNA/pmol coupled dye). They found that optimal fluorescent targets contain >2500 pmol of cDNA per reaction, >125 pmol of incorporated dye per reaction, and <40 nucleotides/dye molecule. Assuming the cDNA synthesis and labeling reactions are successful, 50 pmol of Cy3™ is combined with an equivalent quantity of Cy5™ and evaporated to dryness in a roto-evaporator set to 45°C. The combined dried cDNA targets are then suspended in 70 μl of hybridisation solution.

REFERENCES

Bishop M. J., (Editor) *Genetics Databases*. Academic Press. 1999.

Mount D. W., *Bioinformatics: Sequence and Genome Analysis*, Cold Spring Harbor Laboratory Press, 2001.

Sensen C., *Essentials of Genomics and Bioinformatics*, John Wiley & Sons. 2002.

6

Gene Mapping and Prediction

Genome mapping is the creation of a genetic map assigning DNA fragments to chromosomes. When a genome is first investigated, this map is non existent. The map improves with the scientific progress and is perfect when the genomic DNA sequencing of the species has been completed. During this process, and for the investigation of differences in strain, the fragments are identified by small tags. These may be genetic markers (PCR products) or the unique sequence-dependent pattern of DNA-cutting enzymes. The ordering is derived from genetic observations (recombinant frequency) for these markers or in the second case from a computational integration of the fingerprinting data. The term "mapping" is used in two different but related contexts.

Two different ways of mapping are distinguished. Genetic mapping uses classical genetic techniques (e.g. pedigree analysis or breeding experiments) to determine sequence features within a genome. Using modern molecular biology techniques for the same purpose is usually referred to as physical mapping.

MAPPING ELEMENTS OF GENOME

DNA Markers

A DNA marker is simply a uniquely identifiable segment

of DNA. There are several different types of markers, usually ranging in size from one to 300-400 nucleotide bases in size. Markers can be thought of as landmarks, and a set of markers whose relative positions (or order) within a genome are known comprises a map. Markers can be categorised in several ways. Some markers are polymorphic, and others are not (monomorphic). Detection of markers may be either PCR based or hybridisation based. Some markers lie in a sequence of DNA that is expressed; some do not, or their expression status may be unknown.

PCR-based markers are commonly referred to as sequence-tagged sites (STSs). An STS is defined as a segment of genomic DNA that can be uniquely PCR amplified by its primer sequences. STSs are commonly used in the construction of physical maps. STS markers may be developed from any genomic sequence of interest, such as from characterised and sequenced genes, or from expressed sequence tags. Alternatively, STSs may be randomly identified from total genomic DNA. The EST database (dbEST) at NCBI stores information on most STS markers.

Polymorphic Markers

Polymorphic markers are those that show sequence variation among individuals. Polymorphic markers are used to construct genetic linkage maps. The number of alleles observed in a population for a given polymorphism, which can vary from two to >30, determines the degree of polymorphism. For many studies, highly polymorphic markers (>5 alleles) are most useful.

Polymorphisms may arise from several types of sequence variations. One of the earlier types of polymorphic markers used for genomic mapping is a restriction fragment length polymorphism (RFLP). An RFLP arises from changes in the sequence of a restriction enzyme recognition site, which alters the digestion patterns observed during hybridisation-based analysis. Another type of hybridisationbased marker arises from a variable number of tandem repeat units (VNTR). A VNTR locus usually has several alleles, each containing a different number of copies of a common motif of at least 16 nucleotides tandemly oriented along a chromosome.

A third type of polymorphism is due to tandem repeats of short sequences that can be detected by PCR-based analysis. These are known variously as microsatellites, short tandem repeats (STRs), STR polymorphisms (STRPs), or short sequence length polymorphisms (SSLPs). These repeat sequences usually consist of two, three, or four nucleotides and are plentiful in most organisms. All PCR-converted STR markers (those for which a pair of oligonucleotides flanking the polymorphic site suitable for PCR amplification of the locus has been designed) are considered to be STSs. The advent of PCR-based analysis quickly made microsatellites the markers of choice for mapping.

Another polymorphic type of PCR-based marker is a single nucleotide polymorphism (SNP), which results from a base variation at a single nucleotide position. Most SNPs have only two alleles (biallelic). Because of their low heterozygosity, maps of SNPs require a much higher marker density than maps of microsatellites. SNPs occur frequently in most genomes, with one SNP occurring on average approximately once in every 100-300 bases in humans. SNPs lend themselves to highly automated fluidic or DNA chip-based analyses and have quickly become the focus of several large-scale development and mapping projects in humans and other organisms. Further details about all of these types of markers can be found elsewhere.

DNA Clones

The possibility of physically mapping eukaryotic genomes was largely realised with the advent of cloning vehicles that could efficiently and reproducibly propagate large DNA fragments. The first generation of large-insert cloning was made possible with yeast artificial chromosome (YAC) libraries. Because YACs can contain fragments up to 2 Mb, they are suitable for quickly making low-resolution maps of large chromosomal regions, and the first whole-genome physical maps of several eukaryotes were constructed with YACs. However, although YAC libraries work well for ordering STSs and for joining small physical maps, the high rate of chimerism and instability of these clones makes them unsuitable for DNA sequencing.

The second and current generation of large-insert clones consists of bacterial artificial chromosomes (BACs) and P1-artificial chromosomes, both of which act as episomes in bacterial cells rather than as eukaryotic artificial chromosomes. Bacterial propagation has several advantages, including higher DNA yields, ease-of-use for sequencing, and high integrity of the insert during propagation. As such, despite the relatively limited insert sizes (usually 100-300 kb), BACs and PACs have largely replaced YACs as the clones of choice for large-genome mapping and sequencing projects. DNA fingerprinting has been applied to BACs and PACs to determine insert overlaps and to construct clone contigs.

In this technique, clones are digested with a restriction enzyme, and the resulting fragment patterns are compared between clones to identify those sharing subsets of identically sized fragments. In addition, the ends of BAC and PAC inserts can be directly sequenced; clones whose insert-end sequences have been determined are referred to as sequence-tagged clones (STCs). Both DNA fingerprinting and STC generation now play instrumental roles in physical mapping strategies.

MAPPING STRATEGIES

Cytogenetic Maps

Cytogenetic maps are those in which the markers are

localised to chromosomes in a manner that can be directly imaged. Traditional cytogenetic mapping hybridises a radioactively or fluorescently labeled DNA probe to a chromosome preparation, usually in parallel with a chromosomal stain such as Giemsa, which produces a banded karyotype of each chromosome. This allows assignment of the probe to a specific chromosomal band or region. Assignment of cytogenetic positions in this manner is dependent on some subjective criteria (variability in technology, methodology, interpretation, reproducibility, and definition of band boundaries). Thus, inferred cytogenetic positions are often fairly large and occasionally overinterpreted, and some independent verification of cytogenetic position determinations is warranted for crucial genes, markers, or regions.

Probes used for cytogenetic mapping are usually large-insert clones containing a gene or polymorphic marker of interest. Despite the subjective aspects of cytogenetic methodology, karyotype analysis is an important and relatively simple clinical genetic tool; thus, cytogenetic positioning remains an important parameter for defining genes, disease loci, and chromosomal rearrangements. Newer cytogenetic techniques such as interphase fluorescence in situ hybridisation (FISH) and fiber FISH instead examine chromosomal preparations in which the DNA is either naturally or mechanically extended.

Studies of such extended chromatin have demonstrated a directly proportional relationship between the distances measured on the image and the actual physical distance for short stretches, so that a physical distance between two closely linked probes can be determined with some precision. However, these techniques have a limited ordering range ($\leq$ 1-2 Mb) and are not well-suited for high-throughput mapping.

Genetic Linkage Maps

Genetic linkage (GL) maps (also called meiotic maps) rely

on the naturally occurring process of recombination for determination of the relative order of, and map distances between, polymorphic markers. Crossover and recombination events take place during meiosis and allow rearrangement of genetic material between homologous chromosomes. The likelihood of recombination between markers is evaluated using genotypes observed in multigenerational families. Markers between which only a few recombination occur are said to be linked, and such markers are usually located close to each other on the same chromosome. Markers between which many recombinations take place are unlinked and usually lie far apart, either at opposite ends of the same chromosome or on different chromosomes.

Because the recombination events cannot be easily quantified, a statistical method of maximum likelihood is usually applied in which the likelihood of two markers being linked is compared with the likelihood of being unlinked. This likelihood ratio is called a "lod" score (for "log of the odds"), and a lod score greater than 3 (corresponding to odds of 1,000:1 or greater) is usually taken as evidence that markers are linked. The lod score is computed at a range of recombination fraction values between markers (from 0 to 0.5), and the recombination fraction at which the lod score is maximised provides an estimate of the distance between markers.

A map function is then used to convert the recombination fraction into an additive unit of distance measured in centiMorgans (cM), with 1 cM representing a 1% probability that a recombination has occurred between two markers on a single chromosome. Because recombination events are not randomly distributed, map distances on linkage maps are not directly proportional to physical distances.

The majority of linkage maps are constructed using multipoint linkage analysis, although multiple pairwise linkage analysis and minimisation of recombination are also valid approaches. Commonly used and publicly available computer programs for building linkage maps include LINKAGE, CRI-MAP, MultiMap, MAPMAKER, and MAP. The MAP-O-MAT Web server is available for estimation of map distances and for evaluation of statistical support for order.

Because linkage mapping is a based on statistical methods, linkage maps are not guaranteed to show the correct order of markers. Therefore, it is important to be critical of the various available maps and to be aware of the statistical criteria that were used in map construction. Typically, only a subset of markers is mapped with high statistical support. The remainder are either placed into well-supported intervals or bins or placed into unique map positions but with low statistical support for order. To facilitate global coordination of human linkage mapping, DNAs from a set of reference pedigrees collected for map construction were prepared and distributed by the Centre d'Etude du Polymorphism Humain.

Nearly all human linkage maps are based on genotypes from the CEPH reference pedigrees, and genotypes for markers scored in the CEPH pedigrees are deposited in a public database maintained at CEPH. Most recent maps are composed almost entirely of highly polymorphic STR markers. These linkage maps have already exceeded the maximum map resolution possible given the subset of CEPH pedigrees that are commonly used for map construction, and no further large-scale efforts to place STR markers on human linkage maps are planned. Thousands of SNPs are currently being identified and characterised, and a subset are being placed on linkage maps. Linkage mapping is also an important tool in experimental animals, with many maps already produced at high resolution and others still under development.

Radiation Hybrid Maps

Radiation hybrid (RH) mapping is very similar to linkage mapping. Both methods rely on the identification of chromosome breakage and reassortment. The primary difference is the mechanism of chromosome breakage. In the construction of radiation hybrids, breaks are induced by the application of lethal doses of radiation to a donor cell line, which is then rescued by fusion with a recipient cell line (typically mouse or hamster) and grown in a selective medium such that only fused cells survive. An RH panel is a library of fusion cells, each of which has a separate collection of donor fragments. The complete donor genome is represented multiple times across most RH panels. Each fusion cell, or radiation hybrid, is then scored by PCR to determine the presence or absence of each marker of interest.

Markers that physically lie near each other will show similar patterns of retention or loss across a panel of RH cells and behave as if they are linked, whereas markers that physically lie far apart will show completely dissimilar patterns and behave as if they are unlinked. Because the breaks are largely randomly distributed, the break frequencies are roughly directly proportional to physical distances. The resulting data set is a series of positive and negative PCR scores for each marker across the hybrid panel. These data can be used to statistically infer the position of chromosomal breaks, and, from that point on, the procedures for map construction are similar to those used in linkage mapping. A map function is used to convert estimates of breakage frequency to additive units of distance measured in centirays (cR), with 1 cR representing a 1% probability that a chromosomal break has occurred between two markers in a single hybrid.

The resolution of a radiation hybrid map depends on the size of the chromosomal fragments contained in the hybrids, which in turn is proportional to the amount of irradiation to which the human cell line was exposed. Most RH maps are built using multipoint linkage analysis, although multiplepairwise linkage analysis and minimisation of recombination are also valid approaches. Three genome-wide RH panels exist for humans and are commercially available, and RH panels are available for many other species as well.

Widely used computer programs for RH mapping are RHMAP, RHMAPPER, and MultiMap, and on-line servers that allow researchers to place their RH mapped markers on existing RH maps are available. The Radiation Hybrid Database (RHdb) is the central repository for RH data on panels available in all species. The Radiation Hybrid Information Web site also contains multi-species information about available RH panels, maps, ongoing projects, and available computer programs.

Transcript Maps

Of particular interest to researchers chasing disease genes are maps of transcribed sequences. Although the transcript sequences are mapped using one of the methods described in this section, and thus do not require a separate mapping technology, they are often set apart as a separate type of map. These maps consist of expressed sequences and sequences derived from known genes that have been converted into STSs and usually placed on conventional physical maps. Recent projects for creating large numbers of ESTs have made tens of thousands of unique expressed sequences available to the mapping laboratories. Transcribed sequence maps can significantly speed the search for candidate genes once a disease locus has been identified. The largest human transcript map to date is the GeneMap '99, described below.

Physical Maps

Physical maps include maps that either are capable of

directly measuring distances between genomic elements or that use cloned DNA fragments to directly order elements. Many techniques have been created to develop physical maps. The most widely adopted methodology, due largely to its relative simplicity, is STS content mapping. This technique can resolve regions much larger than 1 Mb and has the advantage of using convenient PCR-based positional markers.

In STS content maps, STS markers are assayed by PCR against a library of large-insert clones. If two or more STSs are found to be contained in the same clone, chances are high that those markers are located close together. The STS content mapping technique builds a series of contigs (i.e., overlapping clusters of clones joined together by shared STSs). The resolution and coverage of such a map are determined by a number of factors, including the density of STSs, the size of the clones, and the depth of the clone library. Maps that use cloning vectors with smaller insert sizes have a higher theoretical resolution but require more STSs to achieve coverage of the same area of the genome. Although it is generally possible to deduce the relative order of markers on STS content maps, the distances between adjacent markers cannot be measured with accuracy without further experimentation, such as by restriction mapping. However, STS content maps have the advantage of being associated with a clone resource that can be used for further studies, including subcloning, DNA sequencing, or transfection.

Several other techniques in addition to STS content and radiation hybrid mapping have also been used to produce physical maps. Clone maps rely on techniques other than STS content to determine the adjacency of clones. For example, the CEPH YAC map used a combination of fingerprinting, inter-Alu product hybridisation, and STS content to create a map of overlapping YAC clones. Fingerprinting is commonly used by sequencing centers to assemble and/or verify BAC and PAC contigs before clones are chosen for sequencing, to select new clones for sequencing that can extend existing contigs, and to help order genomic sequence tracts generated in whole-genome sequencing projects.

Sequencing of large-insert clone ends (STC generation), when applied to a wholegenome clone library of adequate coverage, is very effective for whole-genome mapping when used in combination with fingerprinting of the same library. Deletion and somatic cell hybrid maps relying on large genomic reorganisations (induced deliberately or naturally occurring) to place markers into bins defined by chromosomal breakpoints have been generated for some human chromosomes. Optical mapping visualises and measures the length of single DNA molecules extended and digested with restriction enzymes by high-resolution microscopy. This technique, although still in its infancy, has been successfully used to assemble whole chromosome maps of bacteria and lower eukaryotes and is now being applied to complex genomes.

Comparative Maps

Comparative mapping is the process of identifying conserved chromosome segments across different species. Because of the relatively small number of chromosomal breaks that have occurred during mammalian radiation, the order of genes usually is preserved over large chromosomal segments between related species. Orthologous genes (copies of the same genes from different species) can be identified through DNA sequence homology, and sets of orthologous genes sharing an identical linear order within a chromosomal region in two or more species are used to identify conserved segments and ancient chromosomal breakpoints.

Knowledge about which chromosomal segments are shared and how they have become rearranged over time greatly increases our understanding of the evolution of different plant and animal lineages. One of the most valuable applications of comparative maps is to use an established gene map of one species to predict positions of orthologous genes in another species. Many animal models exist for diseases observed in humans. In some cases, it is easier to identify the responsible genes in an animal model than in humans, and the availability of a good comparative map can simplify the process of identifying the responsible genes in humans. In other cases, more might be known about the gene(s) responsible in humans, and the same comparative map could be used to help identify the gene(s) responsible in the model species. There are several successful examples of comparative candidate gene mapping.

As mapping and sequencing efforts progress in many species, it is becoming possible to identify smaller homologous chromosome segments, and detailed comparative maps are being developed between many different species. Fairly dense gene-based comparative maps now exist between the human, mouse, and rat genomes and also between several agriculturally important mammalian species. Sequence- and protein-based comparative maps are also under development for several lower organisms for which complete sequence is available. A comparative map is typically presented either graphically or in tabular format, with one species designated as the index species and one or more others as comparison species. Homologous regions are presented graphically with nonconsecutive segments from the comparison species shown aligned with their corresponding segments along the map of the index species.

Integrated Maps

Map integration provides interconnectivity between

mapping data generated from two or more different experimental techniques. However, achieving accurate and useful integration is a difficult task. Most of the genomic maps and associated Web sites discussed in this section provide some measure of integration, ranging from the approximate cytogenetic coordinates provided in the Généthon GL map to the interassociated GL, RH, and physical data provided by the Whitehead Institute (WICGR) Web site. Several integration projects have created truly integrated maps by placing genomic elements mapped by differing techniques relative to a single map scale. The most advanced sources of genomic information provide some level of genomic cataloguing, where considerable effort is made to collect, organise, and map all available positional information for a given genome.

DIFFICULTIES IN MAPPING

It is important to realise that the genomic mapping information currently available is a collection of a large number of individual data sets, each of which has unique characteristics. The experimental techniques, methods of data collection, annotation, presentation, and quality of the data differ considerably among these data sets. Although most mapping projects include procedures to detect and eliminate and/or correct errors, there are invariably some errors that occur, which often result in the incorrect ordering or labeling of individual markers. Although the error rate is usually very low (5% or less), a marker misplacement can obviously have a great impact on a study.

A few mapping Web sites are beginning to flag and correct (or at least warn) users of potential errors, but most errors cannot be easily detected. Successful strategies for minimising the effects of data error include (1) simultaneously assessing as many different maps as possible to maximise redundancy; (2) increased emphasis on utilising integrated maps and genomic catalogues that provide access to all available genomic information for the region of interest; and (3) if possible, experimentally verifying the most critical marker positions or placements.

In addition to data errors, several other, more subtle complexities are notable. Foremost is the issue of nomenclature, or the naming of genomic markers and elements. Many markers have multiple names, and keeping track of all the names is a major bioinformatics challenge. For example, the polymorphic marker D1S243 has several assigned names: AFM214yg7, which is actually the name of the DNA clone from which this polymorphism was identified; SHGC-428 and stSG729, two examples of genome centers renaming a marker to fit their own nomenclature schemes; and both GDB:201358 and GDB:133491, which are database identifier numbers used to track the polymorphism and STS associated with this marker, respectively, in the Genome Database (GDB).

Genomic mapping groups working with a particular marker often assign an additional name to simplify their own data management, but, too often, these alternate identifiers are subsequently used as a primary name. Furthermore, many genomic maps display only one or a few names, making comparisons of maps problematic. Mapping groups and Web sites are beginning to address these inherent problems, but the difficulty of precisely defining "markers," "genes," and "genomic elements" adds to the confusion. It is important to distinguish between groups of names defining different elements.

A gene can have several names, and it can also be associated with one or more EST clusters, polymorphisms, and STSs. Genes spanning a large genomic stretch can even be represented by several markers that individually map to different positions. Web sites providing genomic cataloguing, such as LocusLink, UniGene, GDB,

GeneCards, and eGenome, list most names associated with a given genomic element. Nevertheless, collecting, cross-referencing, and frequently updating one's own sets of names for markers of interest is also a good practice, as even the genomic cataloguing sites do not always provide complete nomenclature collections. Each mapping technique yields its own resolution limits.

Cytogenetic banding potentially orders markers separated by ≥1-2 Mb, and genetic linkage (GL) and RH analyses yields long-range resolutions of ≥0.5-1 Mb, although localised ordering can achieve higher resolutions. The confidence level with which markers are ordered on statistically based maps is often overlooked, but this is crucial for assessing map quality. For genomes with abundant mapping data such as human or mouse, the number of markers used for mapping often far exceeds the ability of the technique to order all markers with high confidence. Mappers have taken two approaches to address this issue. The first is to order all markers in the best possible linear order, regardless of the confidence for map position of each marker.

Alternatively, the high confidence linear order of a subset of markers is determined, and the remaining markers are then placed in high confidence "intervals," or regional positions. The advantage of the first approach is that resolution is maximised, but it is important to pay attention to the odds for placement of individual markers, as alternative local orders are often almost equally likely. Thus, beyond the effective resolving power of a mapping technique, increased resolution often yields decreased accuracy, and researchers are cautioned to strike a healthy balance between the two.

Each mapping technique also yields very different measures of distance. Cytogenetic approaches, with the exception of high-resolution fiber FISH, provide only rough distance estimates, GL and STS content mapping provide marker orientation but only relative distances, and RH mapping yields distances roughly proportional to true physical distance. For GL analysis, unit measurements are in centMiorgans, with 1 cM equivalent to a 1% chance of recombination between two linked markers.

The conversion factor of 1 cM $\simeq$ 1 Mb is often cited for the human genome but is overstated, as this is just the average ratio genome-wide, and many chromosomal regions have recombination hotspots and coldspots in which the cM-to-Mb ratio varies as much as 10-fold. In general, cytogenetic maps provide subband marker regionalisation but limited localised ordering, GL and STS content maps provide excellent ordering and limited-to-moderate distance information, and RH maps provide the best combination of localised ordering and distance estimates.

Finally, there are various levels at which genomic information can be presented. Single-resource maps such as the Généthon GL maps use a single experimental technique and analyse a homogeneous set of markers. Strictly comparative maps make comparisons between two or more different single-dimension maps either within or between species but without combining data sets for integration. GDB's Mapview program can display multiple maps in this fashion.

Integrated maps recalculate or completely integrate multiple data sets to display the map position of all genomic elements relative to a single scale; GDB's Comprehensive Maps are an example of such integration. Lastly, genome cataloguing is a relatively new way to display genomic information, in which many data sets and/or Web sites are integrated to provide a comprehensive listing and/or display of all identified genomic elements for a given chromosome or genome.

Completely sequenced genomes such as *C. elegans* and *S. cerevisiae* have advanced cataloguing efforts, but catalogues for complex genome organisms are in the early stages. Examples include the interconnected NCBI databases, MGD, and eGenome. Catalogues provide a "onestop shopping" solution to collecting and analysing genomic data and are recommended as a maximum-impact means to begin a regional analysis. However, the individual data sets provide the highest quality positional information and are ultimately the most useful for region definition and refinement.

REALISTIC USES OF MAPPING RESOURCES

Potential applications of genomic data are numerous and, to a certain extent, depend on the creativity and imagination of the researcher. However, most researchers utilise genomic information in one of three ways: to find out what genomic elements-usually transcribed elements-are contained within a genomic region, to determine the order of defined elements within a region, or to determine the chromosomal position of a particular element. Each of these goals can be accomplished by various means, and the probability of efficient success is often enhanced by familiarity with many of the resources. It is prudent to follow a logical course when using genomic data.

During the initial data acquisition step, in which genomic data are either generated experimentally or retrieved from publicly available data sources, simultaneous evaluation of multiple data sets will ensure both higher resolution and greater confidence while increasing the likelihood that the genomic elements of interest are represented. Second, the interrelationships and limitations of the data sets must be sufficiently understood, as it is easy to overinterpret or underrepresent the data. Finally, it is important to verify critical assignments independently, especially when using mapping data that are not ordered with high confidence.

Genomic Region

A genomic region of interest is best defined by two flanking markers that are commonly used for mapping purposes, such as polymorphic Ge´ne´thon markers in humans or MIT microsatellites in mice. Starting with a cytogenetically defined region is more difficult due to the subjective nature of defining chromosomal band boundaries. Conversion of cytogenetic boundaries to representative markers can be approximated by viewing the inferred cytogenetic positions of markers in comprehensive maps such as GDB's universal map, UDB, LDB, or eGenome. Because these cytogenetic positions are inferred and approximate, a conservative approach is recommended when using cytogenetic positions for region definition. The choice of flanking markers will impact how precisely a region's size and exact boundary locations can be defined.

Commonly used markers are often present on multiple, independently derived maps, so their "position" on the chromosome provides greater confidence for anchoring a regional endpoint. In contrast, the exact location of less commonly used markers is often locally ambiguous. These markers can sometimes be physically tethered to other markers if a large sequence tract that contains multiple markers can be found. This can be performed by BLASTing marker sequences against GenBank or by scanning e-PCR results in UniGene or eGenome for a particular marker.

Determining and Ordering

Once a region has been defined, there are a number of resources available for determining what lies within the region. A good way to start is to identify a map that contains both flanking markers, either from a chromosome-wide or genome-wide map from the sources listed above, from a genomic catalogue, or from a local map that has been generated by a laboratory interested in this particular region. For humans, GDB is the most inclusive map repository, although many regional maps have not been deposited in GDB but can be found with a literature search of the corresponding cytogenetic band or a gene known to map to the region.

Many localised maps are physically based and are more accurate than their computationally derived, whole-chromosome counterparts. For other species, the number of maps to choose from is usually limited, so it is useful to first define flanking markers known to be contained in the available maps. The map or maps containing the flanking markers can then be used to create a consensus integrated map of the region. This is often an inexact and tedious process. To begin, it is useful to identify from the available maps an index map that contains many markers, high map resolution, and good reliability. Integration of markers from additional maps relative to the index map proceeds by comparing the positions of markers placed on each map.

For example, if an index map contains markers in the order A-B-C-D and a second map has markers in the order B-E-D, then marker E can be localised to the interval between markers B and D on the index map. Importantly, however, the relative position of marker E with respect to marker C usually cannot be accurately determined by this method. Repeated iterations of this process should allow localisation of all markers from multiple maps relative to the index map. This process is of course significantly reinforced by experimental verification, such as with STS content mapping of large-insert clones identified for the regionspecific markers or, ideally, by sequence-determined order.

Each marker represents some type of genomic element: a gene, an EST, a polymorphism, a large-insert clone end, or a random genomic stretch. In humans, identifying what a marker represents is relatively straightforward. Simply search for the marker name in GDB or eGenome, and, in most cases, the resulting Web display will provide a summary of what the marker represents, usually along with hyperlinks to relevant functional information. For mice, MGD provides a similar function to GDB. For other organisms, the best source is usually either dbSTS or, if present, Web sites or publications associated with the underlying maps.

GenBank and dbSTS are alternatives for finding markers, but, because these repositories are passive, many marker sets are not represented. If a marker is known to be expressed, UniGene, LocusLink, and dbEST are excellent sources of additional information. Many genes and some polymorphisms have been independently discovered and developed as markers multiple times, and creating a nonredundant set from a collection of markers is often challenging. GDB, eGenome, MGD, and UniGene are good sources to use for finding whether two markers are considered equivalent but even more reliable is a DNA sequence or sequence contig containing both marker's primers. BLAST and the related BLAST2 are efficient for quickly determining sequence relatedness.

Obviously, the most reliable tool for marker ordering is a DNA sequence or sequence contig. For expressed human markers, searching with the marker name in UniGene or Entrez Genomes returns a page stating where the marker has been mapped in GeneMap '99 and other maps, a list of mRNA, genomic, and EST sequences, and with Entrez Genomes, a Mapviewer-based graphical depiction of the maps, sequence-ready contigs, and available sequence of the region. Similarly, GDB and eGenome show which DNA sequences contain each displayed marker. For other markers, the sequence from which the marker is derived, or alternatively one of the primer sequences, may be used to perform a BLAST search that can identify completely or nearly homologous sequences.

The nonredundant, EST, GSS, and HTGS divisions of GenBank are all potentially relevant sources of matching sequence, depending on the aim of the project. Only long sequences are likely to have worthwhile marker-ordering capabilities. Finished genomic sequence tracts have at least some degree of annotation, and scanning the GenBank record for the large sequence will often yield an annotated list of what markers lie within the sequence and where they are. Keep in mind that such annotations vary considerably in their thoroughness and most are fixed in time; that is, they only recognise markers that were known at the time of the annotation. BLAST, BLAST2, or other sequence-alignment programs are helpful in identification or confirmation of what might lie in a large sequence. Also, the NCBI e-PCR Web interface can be used to identify all markers in dbSTS contained within a given sequence, and this program can be installed locally to query customised marker sets with DNA sequences.

For genomes for which DNA sequencing is complete or is substantially underway, it may be possible to construct local clone or sequence contigs. Among higher organisms, this is currently possible only for the human and mouse genomes. Although individual clone sequences can be found in GenBank, larger sequence contigs -sequence tracts comprising more than one BAC or PAC-are more accessible using the Entrez Genomes Web site. Here, by entering a marker or DNA accession number into the contigs search box, researchers can identify sequence contigs containing that marker or element. This site also provides a graphical view of all other markers contained in that sequence, the base pair position of the markers in the sequence, and, with the Mapviewer utility, graphical representations of clone contigs. This process can also be performed using BLAST or e-PCR, although it is somewhat more laborious.

Once a sequence has been identified for markers in a given region, YAC clone, DNA fingerprinting, and STC data can be used to bridge gaps. For humans and mice, the WICGR YAC data provide a mechanism for identifying YAC clones linking adjacent markers. However, caution should be exercised to rely mainly on doublelinked contigs and/or to experimentally confirm YAC/marker links. Also for human genome regions, the UWHTSC and TIGR Web sites for identifying STCs from DNA sequence or BAC clones are very useful. For example, researchers with a sequence tract can go to the UWHTSC TSC search page, enter their sequence, and find STCs contained in the sequence. Any listed STC represents the end of a BAC clone whose insert contains a portion of the input sequence.

The TIGR search tool is complementary to the UWHTSC search, as the TIGR site requires input of a large-insert clone name, which yields STC sequences. STCs represent large-insert clones that potentially extend a contig or link two adjacent, nonoverlapping contigs. Similarly, ≥375,000 human BAC clones have been fingerprinted for rapid identifi-cation of overlapping clones. The fingerprinting data are available for searching at the Washington University Human Genome Sequencing Center (WUGSC). Combined use of Entrez, BLAST, the STC resources, and the BAC fingerprinting data can often provide quick and reliable contig assembly by in silico sequence and clone walking.

Gene Prediction

Gene finding typically refers to the area of computational biology that is concerned with algorithmically identifying stretches of sequence, usually genomic DNA, that are biologically functional. This especially includes protein-coding genes, but may also include other functional elements such as RNA genes and regulatory regions. Gene finding is one of the first and most important steps in understanding the genome of a species once it has been sequenced.

In its earliest days, "gene finding" was based on painstaking experimentation on living cells and organisms. Statistical analysis of the rates of homologous recombination of several different genes could determine their order on a certain chromosome, and information from many such experiments could be combined to create a genetic map specifying the rough location of known genes relative to each other. Today, with comprehensive genome sequence and powerful computational resources at the disposal of the research community, gene finding has been redefined as a largely computational problem.

Determining that a sequence *is functional* should be distinguished from determining *the function* of the gene or its product. The latter still demands *in vivo* experimentation through gene knockout and other assays, although frontiers of bioinformatics research are making it increasingly possible to predict the function of a gene based on its sequence alone.

There are two types of genetic elements with respect to primary structure of proteins: short cis-regulatory elements that control how the gene is expressed, and coding sequences (CDSs) that code for the gene product (protein). Since the majority of the mammalian genes have introns, the key in gene prediction is to detect splice site signals and to locate CDSs.

In the early 80s, computer methods were developed to find genes either by detecting extended CDS regions based on codon usage or coding periodicity , or by detecting splicing signals. Later, more complex algorithms based on modern statistical, linguistic or machine-learning techniques have been developed to combine various sequence features for more accurate gene predictions. As more genomes become available, comparative genomics can offer even better prediction than ab initio methods for homologous genes. Although gene prediction tools have become more sophisticated, prediction accuracy is still far from satisfactory. Up to now, the total number of human genes, ranging from 24,500 to 45,000, still cannot be estimated with certainty, and current mammalian gene counts are highly hypothetical in nature. There is still a long way to go before gene prediction programs can produce accurate predictions of the gene content in mammalian genomes.

METHODS FOR PREDICTING *AB INITIO* GENES

As most mammalian genes contain introns, there are some basic methods for predicting coding exons of the intron containing genes.

Signals for Exon Prediction

There are four basic signals for any coding exon predictions: the translational start site, (the START site) the 5' splice site, (5'ss, or donor site), the 3' splice site (3'ss, or acceptor site), the branch site and the translational stop site (the STOP codon). The simplest measure for a signal site is so-called consensus (word pattern). For example, a STOP site may be any of TAA, TAG or TGA (the ratio among these three stop codons in mammalian genomes is about 1:1:2). The ATG site and the START site may be characterised by the Kozak consensus GCCGCCRCCATGG.

Splice site consensus was first catalogued by Mount and later was refined with more data by Senapathy et al: donor site AG | GTRAGT, acceptor site (Y)nNCAG | G and branch site CTRAY (the average distance between the branch point "A" to the 3'ss boundary, indicated by " | " in the acceptor site) is 26 bases. Senapathy et al. actually presented the splice signals by more quantitative frequency matrices, where each matrix element Mij is equal to the frequency count of the base "i" at position "j" from the training set of aligned splice site sequences. If one defines $f_{ij} = M_{ij}/100$, and fi as the background frequency of nucleotide i, then the popular log-odd scoring matrix (weight matrix model, or WMM) $S_{ij} = \log(f_{ij} / f_i)$ scores the signal as the sum of the scores over the bases within the signal.

Although useful in many cases, there are at least two main problems with any WMM approache. Firstly, it assumes that bases at different positions are independent. There are many ways to incorporate base dependencies. One method is to assume that each base is correlated with its neighboring bases. It is equivalent to extending the rows to doublets or to triplets. But more dependencies result in more parameters to be estimated and it requires much more training data.

Another method is to apply a decision tree to partition total training data into subsets so that splice site bases within each subset are approximately independent and hence can be modeled by a separate WMM. This also ignores the GC-content of the gene locus. It is well known that a mammalian genome has large variations in GC-content, often referred to as isochores, therefore all signal compositions will be highly biased by the GC-content. More modern gene prediction tools use GC-content specific signal models.

There are other ways one may further improve the accuracy of a signal prediction. For instance, one could model secondary structures of pre-mRNAs. But the most effective way of improving signal site prediction is to combine flanking sequence features. Recently, Pertea et al compared several leading prediction programs: NetGene2, HSPL, NNSplice GENIO and SpliceView with their own GeneSplicer, and concluded that "Overall, NetGene2 appears to be the best for donor site prediction, while for acceptor sites either GeneSplicer, NetGene2 or HSPL perform comparably. One advantage of GeneSplicer for the latter task is that its thresholds can be adjusted by the user to vary the false negative and false positive rates". The best average of error (false-positive + false-negative) rate for either donor or acceptor site prediction is about 5%. This may sound not bad if the search is restricted in a short region. If one uses any predictor to search a large region of a genome, the false-positive rate is unacceptable because for every true site there are hundreds of pseudo-sites.

For example, if a large region has 40 true sites and 4000 pseudo-sites, one true site would be missed (2.5% falsenegatives) and 100 pseudo-sites would be predicted as true sites (2.5% false-positives)! Since adjacent donor site and acceptor site are not independent, this correlation can be explored for further eliminating false-positives. For short introns, occurring mostly in lower eukaryotes, an intron is recognised by the interaction of splicing factors binding across the intron-ends.

In vertebrates, exons are much shorter, recognition of exons by the interaction of splicing factors binding across the exon-ends is the key. Therefore mammalian functional splice sites can only be effectively identified simultaneously through exon recognition. An additional complication, dealt poorly with by current splice site prediction programs is the presence of non-canonical sites. At least 1% of all introns do not conform to the canonical AG-GT boundaries, and they are systematicall ignored by splice site and gene prediction methods.

Prediction of Exons

In addition to signal features, prediction of exons should also require content features for both exon regions and flanking intron regions (typically < 100 bp on each side). To discriminate CDS from introns, the best content feature is the framespecific hexamer frequency score, named so because it captures codon-bias information and codon-codon correlation. There are many ways to construct such a score, both Log-Odd score $L_E(w,i) = \log [f_E(w,i) / f_I(w)]$ and preference score $P_E(w,i) = f_E(w,i) / [f_E(w,i) + f_I(w)]$ are popular among exon finders, where fE(w,i) is the frequency of hexamer *w* in frame *i* calculated from known exon training data and $f_I(w)$ is the frequency of *w* calculated from known flanking introns. If the hexamer *w* is more likely to be found in an exon at the given frame than in an intron, L_E will be positive (P_E will be greater than ½). P_E is easier to use as it ranges from 0 to 1, and has been used in both (linear or quadratic discriminant analysis, LDA/QDA) LDA and QDA exon finders. In addition to coding measures, exon size is another important feature variable one must consider.

For human internal coding exons, the size distribution is close to a log-normal distribution centered around 125 bp. For intron regions, one may also construct similar hexamer scores: $L_I(w) = \log [f_I(w) / f_E(w)] = -L_E(w)$ or $P_I(w) = f_I(w) / [f_E(w) + f_I(w)] = 1 - P_E(w)$, where $f_E(w)$ is the average of $f_E(w,i)$ over all three reading frames (i.e. $f_E(w) = [f_E(w,0) + f_E(w,1) + f_E(w,2)]/3$). These coding measures can only help with prediction of CDS regions, there has not been any effective method for predicting non-translated regions (UTRs) of an exon.

With a training set of exons and pseudo-exons (randomly selected AG-ORF-GT regions, also called a spliceable open reading frame, i.e. open reading frames bounded by the conserved 3'ss AG and 5'ss GT pair) and a set of feature variables $x = (x_1, x_2, .., x_K)$ (for example, x_1 = 5'-flanking intron region score = average LI(w) over all hexamers in the 5'-flanking intron region, x_2 = acceptor site score, x_3 = maximum exon score = average $L_E(w,i)$ over all hexamers in reading frame *i* and then take the maximum over *i*, x_4 = exon size, x_5 = donor site score, x_6 = 3'-flanking intron region score, etc.), each training sample (a true exon or a pseudo-exon) can be represented by a point in the *k*-dimensional feature space.

There are many statistical or machine-learning methods (most of them have been tried) that can be used to build an optimal (in the sense of minimisation of false-positive and false-negative errors using cross-validation tests) discrimination function. This function is a decision surface (exon predictor) in the *k*-dimensional feature space that can best separate true exons from pseudo-exons. In MZEF, QDA is implemented. Namely, the two training sample points (true exons and pseudo-exons) in the feature space are approximated by two separate k-dimensional Gaussian distributions characterised by the sample means and covariance matrices, the intersection of the two Gaussian distributions is a *(k-1)*-dimensional quadratic discrimination surface separating the two sample points in an optimal way. When the two Gaussians are assumed to have the same covariance matrix, the surface will become a hyper-plane and QDA will reduce to LDA, a method implemented in HEXON that is the exonfinder part of FGENEH.

Splice site identification helps exon recognition, internal coding exon measure can help further improve functional splice site selection. Indeed, Thanaraj and Robinson found that (i) a high proportion of false positive splice sites from computational predictions occur in the vicinity of real splice sites; and (ii) current algorithms are misled to predict wrong splice sites more often when the coding potential ends within +/-25 nucleotides from real sites than when it ends at father positions. Their integrated system (MZEF-SPC) with Splice ProximalCheck (SPC) as a front-end filter for MZEF was able to eliminate 2/3 of the predicted false positives at the expense of loosing 1/10 of predicted true positives. In fact, these "false positive" splice sites could be alternate splice sites in a lot of cases. Exon extensions and truncations are commonly seen when an exon is skipped. Existing alternative exons obviously makes it so much harder to predict accurately the splice sites, as there are quite a number of real splice sites. Over half of all genes are alternatively spliced, and over 10 isoforms on average per gene, so the in-silico "false positives" may actually be "false negatives" in-vitro!

Exon Assembly

As exons are not independent, by splicing exons together to assemble a gene one can further eliminate false exon predictions by imposing translatability (i.e. adjacent exons must maintain the open reading frame). The main difficulty in exon assembly is the combinatorial explosion problem: the number of ways N candidate exons may be combined grows exponentially with N. The key idea of computational feasibility comes from dynamic programming (DP), which allows finding "optimal assembly" quickly without having to enumerate all possibilities. DP is also used in GeneParse to recursively search for exon-intron boundaries with signal and content measures obtained by a neural network. The FGENEH algorithm incorporate 5'-, internal, and 3'-exon identification linear discriminant functions and a dynamic programming approach for exon assembly. A more efficient dynamic programming algorithm was introduced by Guigó, and it is incorporated in the program GENEID.

A novel advance in gene prediction methodologies was the application of generalised Hidden Markov Models (HMMs) initially implemented in the Genie algorithm. Soon after, it was implemented in the Genscan algorithm to predict multiple genes. Several other HMM-based gene prediction programs were developed later: VEIL, HMMGENE and FGENESH. In a HMM approach, different types of structure components (such as, exon or intron) are characterised by a state, a gene model is thought to be generated by a state machine: starting from 5' to 3', each base-pair is generated by a "emission probability" conditioned on the current state and surrounding sequences and transition from one state to another is governed by a "transition probability" which obeys all the constraints (such as an intron can only follow an exon, reading frames of two adjacent exons must be compatible, etc.).

All the parameters of the "emission probabilities" and the "transition probabilities" are learned (pre-computed) from some training data. Since the states are unknown ("Hidden"), an efficient algorithm (called the Viterbi algorithm, similar to DP) may be used to select a best set of consecutive states (called a "parse"), which has the highest overall probability compared with any other possible parse of the given genomic sequence. The reason these fully probabilistic state models have become preferable is that all scores are probabilities themselves and the weighting problem has become a matter of counting relative observed state frequency.

Gene Prediction Programs

There are number of the most popular gene finders available on the internet, as well as other useful gene prediction resources. It has been reported that by integrating different programs together, one could achieve higher accuracy. For example, in DIGIT, a Bayesian procedure and a HMM were used to combine the results of FGENESH, GENSCAN and HMMGENE to report better performance than the individual programs.

REFERENCES

Marashi SA, Eslahchi C, Pezeshk H, Sadeghi M., "Impact of RNA structure on the prediction of donor and acceptor splice sites", *BMC Bioinformatics,* Entrez PubMed, 2006.

Saeys Y, Rouzé P, Van de Peer Y., "In search of the small ones: improved prediction of short exons in vertebrates, plants, fungi and protists", Bioinformatics, 2007.

7

Protein Microarrays

A **protein** microarray is a piece of glass on which different molecules of protein have been affixed at separate locations in an ordered manner thus forming a microscopic array. These are used to identify protein-protein interactions, to identify the substrates of protein kinases, or to identify the targets of biologically active small molecules. The most common protein microarray is the antibody microarray, where antibodies are spotted onto the protein chip and are used as *capture molecules* to detect proteins from cell lysate solutions.

Related microarray technologies also include DNA microarrays, Antibody microarrays, Tissue microarrays and Chemical Compound Microarrays.

TECHNOLOGICAL FOUNDATIONS

Nucleic acid microarrays primarily use short oligonucleotides (15-25 nt), long oligonucleotides (50-120 nt) and PCR-amplified cDNAs (100-3,000 base pairs) as array elements. Genotyping applications, such as those designed to detect single-nucleotide polymorphisms (SNPs), require single-mismatch discrimination and use short oligonucleotides (oligos) to maximise the effect of chemical destabilisation caused by mispairing. Short oligonucleotides and cDNAs also work well for expression analysis, although both types of elements have shortcomings. Short oligonucleotides sometimes lack the specificity required to ensure single-gene specificity in complex hybridisations, and for this reason a 'tiling approach' with multiple short oligos per gene is often used. Because of their extended length, PCR PCR amplified cDNAs produce strong signals and high specificity. The cDNA inserts are obtained readily from cDNA libraries, and are typically used for organisms for which there is limited genomic sequence available, although careful attention must be paid during sample preparation to avoid tracking errors.

Long oligonucleotides offer strong hybridisation signals, good specificity, unambiguous sample identification and affordability. They require the availability of genomic sequence information for each organism under study, but the copious amounts of sequence information now flowing from large-scale genome sequencing projects should help in this regard. Long oligonucleotides can be synthesised in 96- and 384-well formats, providing array elements optimised for binding specificity, melting temperature, concentration and purity. A large number of vendors provide high-quality long oligos for use in conjunction with microarray printing technologies. As an alternative, long oligonucleotides can also be synthesised in situ using ink-jets and micromirrors to control combinatorial synthesis directly on the microarray substrate.

Substrates for microarray analysis must provide stable attachment of array elements, low background signals, homogeneous surface chemistry and high-quality data. Glass (silicon dioxide) is the most common substrate material, and a plethora of glass slides and substrates are available, including those with atomically flat surfaces. Typically, reactive amine, aldehyde or epoxide groups are attached covalently to silicon dioxide molecules on the glass surface using silane chemistry, and such surfaces allow stable attachment of oligonucleotides, cDNAs, proteins and other molecules.

Microarray manufacturing technologies can be divided into two main categories: synthesis and delivery. The synthesis approaches build DNA sequences directly on the microarray substrate by joining nucleotide building blocks in succession to produce the array elements. Photolithography, micromirrors, ink-jets and other technologies are used to deliver light, nucleotides and other reagents to the microarray surface for array element synthesis. The delivery approaches use microarray elements prepared 'offline', depositing them onto the microarray substrate after synthesis using pins, ink-jets and other dispensers. Synthesis technologies are generally used for commercial microarray manufacture, and delivery strategies tend to predominate in research laboratories where microarrays are manufactured 'in house'. Contact and non-contact printing methods impart very high precision, with spot diameters varying by <5-10% from feature to feature.

Clean-room environments with low particle counts, regulated temperature and humidity levels, and controlled static electricity, organic vapours and biological contaminants can improve the quality of microarray manufacture in all settings, ranging from the smallest research laboratories to the largest commercial facilities. Fluorescent probes for expression profiling are typically prepared from total RNA or messenger RNA (mRNA) by reverse transcription, although many different labelling strategies are available. Methods that use T7 RNA polymerase produce large amounts of amplified RNA and are widely used to generate probes from small amounts of sample. Because amplified RNA is produced by linear amplification with T7 polymerase, population skewing and the loss of quantitation are minimal. Control and experimental samples can be labelled separately with fluors that have non-overlapping emission spectra, including cyanine, Alexa, and other fluorescent derivatives. Two samples labelled with different fluors can be hybridised to a single chip to derive absolute and comparative expression information in the two samples.

Post-hybridisation detection techniques, including intensely fluorescent dendrimer reagents, offer some major advantages over direct labelling procedures. Direct labelling generally reduces the efficiency of enzymatic synthesis and can introduce fluor-dependent labelling bias, mandating the use of larger amounts of RNA and 'dye-swapping' strategies to obtain strong signals and avoid experimental artifacts. Unincorporated dyes can also elevate fluorescent background if not removed before hybridisation. Post-hybridisation detection methods can reduce or eliminate these complications and provide cleaner, more reproducible results. The use of 'less invasive' nucleotide analogs, such as aminoallyl derivatives, also obviates some of the complications seen in direct fluorescent labelling.

Microarray detection can be accomplished using a variety of fluorescence detection technologies including instruments that contain confocal optics, photomultiplier tubes (PMTs), and charge-coupled devices (CCD). Detection instruments produce graphical images in standard tagged-image file format (TIFF), which are two-dimensional, 16-bit numerical representations of microarray surfaces, with intensity values ranging from 0 to 65,536 (2^{16}). Numerical values are used to assign an expression value to each gene, and the data are typically saved in a tab-delimited format to allow their import into software programs for data mining and modelling.

Transformed and normalised data are represented and modeled using a variety of software tools, including scatter plots, principal component analysis (PCA), cluster diagrams, self-organising maps (SOMs), neural networks and other algorithms. Although the mathematical and statistical basis of the computational tools is complex, each endeavours to provide functionally relevant relationships between genes and gene products, assign putative function to unknown sequences, identify potential disease markers, elucidate the biochemical basis of drug and hormone action, and so forth.

The experimental aspects of microarray analysis are linked to data extraction, analysis and modelling in the microarray workflow process. Intranets and the Internet, together with relational database warehouses, figure centrally in generating, mining, storing and retrieving microarray data Downloadable software ('shareware') packages are available free of charge to microarray researchers worldwide. Forums on microarray data analysis, such as the Critical Assessment of Microarray Data Analysis (CAMDA), endeavour to provide much-needed bioinformatics standards.

As microarray platforms mature and expand into more research venues, including university core facilities, small research centers, public health laboratories and testing clinics, major trends are emerging in such areas as assay diversification, microarray density, genomic and focused arrays, proteomics and genetic screening. The first microarray paper featured the small mustard plant Arabidopsis thaliana, but the technology quickly spread to studies of yeast, human and mouse. Assay diversification has continued at a brisk pace, with >100 organisms featured to date.

Microarrays from one organism can be used to examine evolutionarily related species, permitting phylogenetic analysis on a genomic scale, and, coupled with DNA sequencing data, can be used for functional genomic analysis of any organism in the biosphere. Diversification of the microarray platform is also occurring at the level of array elements, which are quickly expanding beyond DNA to include cellular proteins, organic molecules, carbohydrates, peptides and nanotube precursors. Microarray density, defined as the number of array elements per square centimeter of substrate, has also increased steadily over the years, permitting the manufacture of microarrays with tens of thousands of array elements and the analysis of entire genomes on single chips. Concomitant with trends toward whole-genome analysis is the use of 'focused microarrays' containing only a few hundred elements. Focused microarrays complement whole-genome studies by providing a highly economical approach to detailed pathway analysis involving carefully selected gene subsets.

Microarray technology is also rapidly coming into use in the field of proteomics. Small reaction volumes conserve expensive protein reagents and allow extremely rapid binding kinetics, and the parallel format permits the simultaneous measurement of binding constants, catalytic activities and other important parameters across all of the proteins in the cell, providing a comprehensive view of protein function.

Advanced coupling chemistries allow efficient protein attachment in mild buffers at neutral pH, and fluorescent labels obviate the need for radioisotopes and improve assay safety. The high concentration of printed protein molecules stabilises protein structure, and sophisticated printing buffers protect proteins against oxidation and permit ambient storage. A wide spectrum of biochemical processes can be studied by microarray, including protein-protein interactions, protein-drug interactions, post-translational modifications and others. Protein-protein interactions that impart the activity of hormones, transcription factors, molecular motors and the like can be examined in the context of all of the proteins expressed in the cell, providing a detailed view of biochemical function. Pharmaceutical compounds that bind directly to cellular proteins and alter their function can be studied on a global scale, improving the efficiency of drug discovery.

Glycosylation, phosphorylation, acylation, myristylation, prenylation and other post-translational modifications can also be studied by protein microarray, allowing a first-hand look at protein modification at the level of the proteome. It is anticipated that protein microarrays will increasingly replace many of the traditional macroscopic protein techniques, such as filter binding, column chromatography and gel-shift assays, in the coming years. The availability of gentle contact printing methods allows protein microarrays to be manufactured readily. Continued advances in surface chemistry, the preparation and purification of cellular proteins on a genomic scale, peptide synthesis, and fluorescent and colourimetric labelling and detection schemes will speed the proliferation of protein microarray assays. All of the basic principles established for DNA microarrays are extendable to protein microarrays, though the tools must be tailored to provide for the inherent differences between nucleic acids and proteins.

Another noteworthy trend is the expansion of labelling and detection techniques. One interesting non-fluorescent labelling and detection technology, based on resonance light scattering (RLS), uses gold and silver nanoparticles of defined diameter as molecular labels. RLS particles emit intense monochromatic light in the presence of white excitation light, and the particles are extremely stable even when exposed to intense light sources. Another emerging labelling and detection scheme exploits conventional colourimetric labels, including enzyme reaction products, which exhibit a high degree of spatial resolution on microarrays printed on glass substrates coated with white membrane materials such as nitrocellulose and related polymers.

Colourimetric labels pave the way for the conversion of numerous traditional enzyme-linked immunosorbent assay (ELISA) formats into microarray tests and the concomitant use of traditional flatbed scanners for signal detection. The small scale of colourimetric microarrays as compared to standard ELISA assays provides much greater information content and considerable cost reductions. Genetic screening is another bright horizon for microarray analysis. In genetic screening applications, DNA samples are examined to identify single-nucleotide changes, deletions and other minor sequence variants in genes that underlie genetic and infectious disease.

Two main types of microarray formats have been developed for genetic screening and diagnostic applications. The first format uses oligonucleotide microarrays to detect sequence variants in a single-patient sample. One advantage of 'single-patient' microarrays is that a large number of sequence variants can be screened in a single test, providing extremely affordable genotyping information on a genomic scale. A disadvantage is that the cost per patient is rather high, because a separate chip is required for each individual tested.

MULTIPATIENT MICROARRAYS

'Multipatient' microarrays offer a cost-effective alternative to oligonucleotide microarrays, particularly in applications in which a large number of patients must be screened for a relatively small number of loci. In the multipatient format, patient samples containing defined gene segments are printed into microarrays at distinct locations, and the microarrays are hybridised with fluorescent oligonucleotides complementary to the gene loci of interest. Normal, carrier and disease genotypes are scored easily because the printed patient samples hybridise to the fluorescent oligonucleotides with different efficiencies depending on the genotype of the patient. The main advantage of multipatient microarrays is the capacity to examine thousands or even tens of thousands of patients on one chip. One disadvantage is that the patient samples must be printed into microarrays before genotyping, adding additional time to the assay as compared to pre-printed oligonucleotide microarrays.

Microarray analysis represents a classic 'precision in-precision out' technology. Experiments based on sound experimental design, optimised protocols, properly designed array elements, pure samples, robust manufacturing and surface chemistry, high-quality scanning, and correctly applied sample tracking, quantification and data mining tools yield highly valuable results. The current state of the art provides ≤ 5-10% variation in signal intensities among replicate array elements on the same microarray, and ~10-30% variation among corresponding array elements on different microarrays.

The variability introduced by the contributions of cell doubling time, differing biology of individual transgenic animals, clonal variation, environmental effects on gene expression and so forth almost always exceeds the technical components of microarray assays. These biological contributions need to be accounted for and dealt with in data analysis, through the use of replicates and downstream validation. Bioinformatics programmes are beginning to encompass confidence measures of the data using analysis of replicate hybridisations. These advances will eventually allow microarray expression data to be used as stand alone metrics of biological function.

The main appeal of microarrays-and the reason the technology is proliferating over competing platforms such as beads, filters, gels, mass spectroscopy and PCR-derives from the inherent attributes of parallelism, miniaturisation and automation, and the accuracy of the data vis-à-vis the sophisticated instruments and software tools that are available to the researcher. The capacity to conduct tens of thousands of gene determinations in a single reaction vastly improves the precision of the data and reduces the cost as compared to that of serial methods. A $500 microarray, yielding gene expression measurements for 10,000 genes, provides highly quantitative data for $0.05 per gene. The affordability of microarrays, the spacesaving aspects of miniaturisation and high throughput conferred by automation, and the widespread availability of commercial microarrays impart clear advantages over non-chip technologies.

Microarrays will continue to transform basic research, both technologically and methodologically. The technical transformation will see the replacement of radioisotopes and other low-tech methods with fluorescent microarray assays and computational approaches to biology. Methodologically, there will be a continued shift in the direction of discovery-based research at the expense of hypothesis driven studies.

Recent successes in benchmarking microarrays by the US Food and Drug Administration suggest broad applications for assessing the safety of food, drugs, vaccines, medical devices and other products of consumer interest. Identification of food-borne bacterial pathogens by microarray would reduce the incidence of food poisoning, illness and death associated with bacterial contamination of meat, seafood, dairy products and other foods. A related use of 'bacto-chips' would be in clinical settings, to establish the identity of organisms in patients admitted to hospitals with systemic bacterial infections. The capacity to type unambiguously all the common bacteria on a single chip within a few hours of sampling would allow high-speed testing in agricultural, manufacturing and clinical settings.

The efficacy and safety of pharmaceuticals is also amenable to microarray analysis, and both genotyping and gene expression assays can be envisioned in clinical trials.

Genotyping microarrays could be used to parse the clinical trials population into responders and non-responders, enhancing the accuracy of drug-testing results, and allowing drugs to be tailored to specific subsets of the population according to clearly identifiable markers in the patient population.

Gene expression microarrays could be used to examine the physiological effects of drug administration, allowing the analysis of pathways and the identification of side reactions in which drugs bind promiscuously to cellular proteins, producing toxic side effects. Promising lead compounds could be optimised at the chemical level to minimise side reactions, using gene expression information as the functional readout. Expression arrays could also be used to profile and test medical devices that emit γ-rays, Xrays, microwaves, radio waves and other forms of electromagnetic radiation to explore interactions between biological tissues and various forms of radiation.

Population-based screening also holds great promise for identifying disease carriers, thereby reducing the incidence of genetic disease. Particularly in the case of common genetic disorders that exhibit carrier frequencies of $\geq$1%, affordable microarray-based genotyping assays could be used to identify disease-causing alleles in the population and reduce the chances of passing along inherited disorders. Both oligonucleotide and multipatient microarrays can be used for patient screening, and the availability of this technology may promote the availability of home testing kits for breast cancer, cystic fibrosis, sickle cell anemia and other common genetic diseases.

In one model, consumers would obtain testing kits from a public health clinic or local pharmacy, self-sample a droplet of blood that contains their genetic material, send the sample to a central microarray testing facility, and enter a private access code to obtain the confidential testing results through the Internet. Armed with accurate genotyping information, patients could then work closely and confidentially with their family physician or health care professional to obtain accurate and up-to-date information on lifestyle issues, drug therapies and family counselling advice to ensure the highest-quality health care vis-à-vis their genetics.

MICROARRAY DATA ANALYSIS

Functional genomics is the study of gene function through parallel expression measurements of a genome. The most common tools used to carry out these measurements include complementary DNA microarrays, oligonucleotide microarrays or serial analysis of gene expression (SAGE). Although there are many protocols and types of system available, the basic technique involves extraction of RNA from biological samples in either normal or interventional states. The RNA (or, in some protocols, isolated messenger RNA) is then copied, while incorporating either fluorescent nucleotides or a tag that is later stained with fluorescence. The labelled RNA is then hybridised to a microarray for a period of time, after which the excess is washed off and the microarray is scanned under laser light.

With oligonucleotide microarrays, for which all probes have been designed to be theoretically similar with regard to hybridisation temperature and binding affinity, each microarray measures a single sample and provides an absolute measurement level for each RNA molecule, although this absolute measurement might not correlate exactly with concentration in terms of micrograms per unit volume. With cDNA microarrays, for which each probe has its own hybridisation characteristic, each microarray measures two samples, and provides a relative measurement level for each RNA molecule. Regardless of the technique, the end result is 4,000-50,000 measurements of gene expression per biological sample. As a complete experiment might involve anywhere up to hundreds of microarrays, the resultant RNA-expression data sets can vary greatly in size.

As the cost of microarrays continues to drop, it is clear that microarrays are becoming more integral to the drug discovery process. In addition to the obvious use of functional genomics in basic research and target discovery, such as finding genes expressed in significantly different patterns across samples, there are many other specific uses in this domain. These include: bio-marker determination, to find genes that correlate with and presage disease progression, but are easier to measure and follow in clinical trials; pharmacology, to determine differences in gene expression in tissues exposed to various doses of compounds; toxicogenomics, to find gene-expression patterns in a model tissue or organism exposed to a compound and their use as early predictors of adverse events in humans; target selectivity, to define a compound by the gene expression pattern it provokes in a target tissue and then compare it with other compounds using these patterns; prognostic tests, to find a set of genes that accurately distinguishes one disease from another; and disease subclass determination, to find multiple subcategories of tumours in a single clinical diagnosis.

Many free and commercial software packages are now available to analyse microarray data sets, although it is still difficult to find a single off-the-shelf software package that answers all functional-genomics questions. As the field is still young, when developing a bioinformatics analysis pipeline, it is more important to have a good understanding of both the biology involved and the analytical techniques rather than having the right software.

NORMALISATION IN MICROARRAY MEASUREMENTS

Before multiple microarray measurements can be integrated into a single analysis, the reported measurements need to be normalised, or modified (possibly corrected) to make them comparable. When microarrays are used to collect gene-expression data in an experiment in which the measurements are made at the same time, with homogeneous populations of similar cells and using a single microarray technology, normalisation might simply be a matter of adjusting the overall brightness of each scanned microarray image, assuming that the quantity of RNA is equal. Other normalisation methods include: using expression levels of 'housekeeping' genes; using assumptions that most genes do not change across experiments; using SPLINES; or other non-linear techniques.

Typically, however, functional-genomics experiments are more complicated. Recently, increasing efforts have been invested in characterising the 'noise' in microarray technology. As a result, it has become increasingly clear that there are several substantial sources of noise in microarray data. Intra- and inter-microarray variations can markedly skew the interpretation of such expression data. First, improving the reliability of expression measurements starts with proper experimental design. For example, microarrays can measure across the genome, including genes with expression that is controlled by hormones, such as growth hormone or cortisol. So, if organ samples are acquired at various times during the day, genes that appear to be differentially expressed might only be reflecting normal circadian physiology. Pooling samples before hybridisation might control for this biological 'noise.'

In addition, scanned hybridisation images need to be inspected for artefacts, such as scratches and bubbles. Measuring replicate microarrays for each biological sample allows the modelling of this technical noise. Most reported expression data have been obtained on relatively homogeneous cell populations. However, when RNA is extracted from whole organs or from tumour biopsies, the sources of variation increase. There is substantial heterogeneity of expression in cell sub-populations in most organs and in many tumours. Failure to account for such variation could lead to over-interpretation or spurious functional gene association.

Microdissection of cell sub-populations is possible only in a minority of the systems of interest. If microarray-based gene expression measurements are to be reliable and economical, both at the level of basic biology and clinical assays, then all of these further sources of noise/variation must be incorporated directly into the analytical tools that interpret these data. A further issue that needs to be addressed is the difference between the two most commonly used microarray technologies: spotted cDNA microarrays, which report differences in gene expression between two samples, and oligonucleotide microarrays, which report absolute expression levels.

Normalisation techniques for one microarray technology might not apply to another, owing to differences in assumptions and the distributions of the output measurements. For example, if we assume that in any given experiment, most genes in a cell do not change in expression and an equal number of genes are up- and down regulated (not always a valid assumption), then differential-expression measurements from spotted arrays might be found to be normally distributed, whereas measurements from oligonucleotide microarrays will not have the same distribution. Expression measurements made across microarray technologies are not directly comparable.

For example, the published microarray measurements from the US National Cancer Institute (NCI) 60-cancer-cell-line panel, from spotted arrays and from oligonucleotide arrays, show poor correlation between measurements. This can be explained by the differences in low-level hybridisation and analysis between the two techniques. To be precise, Affymetrix microarrays contain between 11 and

20 pairs of oligonucleotide probes for a target RNA, for which one of the pair is the reverse complement to an ideally unique 25-mer in the RNA and the other contains a mutated middle base pair and serves as a measure of stray signal. Using the differences between these intensities, the Affymetrix quantitative software judges the reliability of each probe pair and calculates a qualitative and quantitative measurement. Other quantitative methods are available in addition to the Affymetrix software.

Compared with this, cDNA microarrays contain a single probe for each target RNA, and the two biological samples are different colours, so that after hybridisation, the two colours are scanned separately and relative expression is determined by comparing intensities. There is striking non-correlation between Affymetrix quantitative measurements and ratios of intensities from a cDNA microarray, because these two technologies clearly measure expression differently.

Supervised or unsupervised Appraoches in RNA Expression

Current methodologies to analyse RNA-expression data sets can be divided into two categories: supervised approaches, or analysis to determine genes that fit a predetermined pattern; and unsupervised approaches, or analysis to characterise the components of a data set without the a priori input or knowledge of a training signal.

Supervised methods are generally used for two purposes: finding genes with expression levels that are significantly different between groups of samples, and finding genes that accurately predict a characteristic of the sample. Most functional-genomics experiments still typically use only a handful of microarrays (or equivalent technology), with samples measured under two or three conditions, and the application has the goal of finding those genes that are significantly differentially expressed. Significance has been evaluated in many different ways, including parametric and non parametric tests, analysis of variance and many others.

Although it would be an understatement to call the analyses of these smaller sets of microarray data trivial, there are several published techniques that have been used to find genes that have the most different expression levels between a few samples. When determining whether a particular gene is differentially expressed between two samples, there are four characteristics that need to be considered: absolute expression level, or whether the gene is expressed at a high or low level; subtractive degree of change between groups, or the difference in expression levels across samples; fold change between groups, or the ratio of expression levels across samples; and reproducibility of the measurement, or whether samples with similar characteristics have similar amounts of the gene transcript.

These four characteristics are related; for example, genes measured at low amounts of expression often have measurements that are less reliable, which leads to poor reproducibility across samples and high-fold changes that do not adequately describe the actual degree of change. All of the available techniques for comparing two sets of microarray measurements essentially evaluate these four characteristics for each gene in various ways to rank genes that are most significantly different.

For larger data sets, comparing microarrays one pair at a time misses trends that might exist between measurements. There are several published supervised methods that find genes or sets of genes that accurately predict sample characteristics, such as distinguishing one type of cancer from another, or a metastatic tumour from a non-metastatic one. These methods might find individual genes, such as the nearest neighbour approach, and/or multiple genes, such as decision trees, neural networks and support vector machines.

Users of unsupervised methods try to find internal structure or relationships in a data set instead of trying to determine how best to predict a 'correct answer'. Within unsupervised learning, there are three classes of technique: feature determination, or determining genes with interesting properties without specifically looking for a particular a priori pattern, such as principal-components analysis; cluster determination, or determining groups of genes or samples with similar patterns of gene expression, such as nearest neighbour clustering, self-organising maps, *k*-means clustering and one- and two-dimensional hierarchical clustering; and network determination, or determining graphs representing gene-gene or gene-phenotype interactions using Boolean networks, Bayesian Networks and relevance networks.

Dissimilarity Measures and Clustering Methods

It is crucial to make a distinction between dissimilarity measures (also known as 'metrics') and clustering methods. A dissimilarity measure indicates the degree of similarity between two genes. A clustering method builds on these dissimilarity measures to create groups of features with similar patterns. A commonly used dissimilarity measure is Euclidean distance, for which each gene is treated as a point in multidimensional space, each axis is a separate biological sample and the coordinate on each axis is the amount of gene expression in that sample. One disadvantage of Euclidean distance is that if measurements are not normalised, correlation of measurements can be missed, the focus being instead on the overall amount of expression. A second disadvantage is that genes that are negatively associated with each other will be missed.

Negative associations include gene interactions, such as those of tumour-suppressor genes. As an example, the tumour repressor protein p53 acts as a transrepressor of several other genes. This means that the higher the level of p53, the less expression of other genes is expected. The concept of negative interaction is clearly different than the concept of no interaction. Another dissimilarity measure that is commonly used is the Pearson Correlation Coefficient, which is measured between two genes that are treated as vectors of measurements.

The disadvantages in using this measure with gene-expression measurements are: first, it assumes that the measurements are normally distributed, which might not be the case for oligonucleotide-microarray measurements; and second, it assumes that genes interact in the assumed linear model, when in biology, a particular gene might best regulate other genes when in the middle of its own range of expression.

Operationally, this measure is sensitive to outliers, and although techniques such as the Rank Correlation Coefficient deal with these by replacing the measurements with ranks, it is not clear whether eliminating the outliers is ideal—many past discoveries have been found by focusing on the outliers in biology.

A third dissimilarity measure is mutual information, which allows for any possible model of interaction between genes and uses each expression-level measurement equally regardless of the actual value, and is therefore not biased by outliers. However, calculating the mutual information requires using discrete expression measurements (for example, representing the gene as 'high' and 'low', or 'high', 'medium' and 'low', and so on), and the mutual information depends on the number of 'bins' used.

Ideally, this would be performed in a gene-specific manner, but sufficient information about the range of expression of each gene in all tissues has not yet been collected. Furthermore, gene-gene associations with high mutual information might not even be functions in the mathematical sense, and might be difficult to explain biologically.

Applications of Analytical Techniques

Once a dissimilarity measure has been chosen, the appropriate analytical technique can be applied. This section describes the four commonly used unsupervised techniques—hierarchical clustering, self-organising maps, relevance networks and principal-components analysis—and two commonly used supervised techniques -nearest neighbours and support vector machines.

Hierarchical clustering

Hierarchical clustering is a commonly used unsupervised technique that builds clusters of genes with similar patterns of expression. This is done by iteratively grouping together genes that are highly correlated in terms of their expression measurements, then continuing the process on the groups themselves.

Dendrograms are used to visualise the resultant hierarchical clustering. A dendrogram represents all genes as leaves of a large, branching tree. Each branch of the tree links two genes, two branches or one of each. Although construction of the tree is initiated by connecting genes that are most similar to each other, genes added later are connected to the branches that they most resemble. Although each branch links two elements, the overall shape of the tree can sometimes be asymmetric. In visually interpreting dendrograms, it is important to pay attention to the length of the branches. Branches connecting genes or other branches that are similar are drawn with shorter branch lengths. Longer branches represent increasing dissimilarity.

Hierarchical clustering is particularly advantageous in visualising overall similarities in expression patterns observed in an experiment, and because of this, the technique has been used in many publications. The number and size of expression patterns within a data set can be estimated quickly, although the division of the tree into actual clusters is often performed visually. It is important to note the few disadvantages in their use. First, hierarchical clustering ignores negative associations, even when the underlying dissimilarity measure supports them. Negative correlations might be crucial in a particular experiment, as described above, and might be missed. Furthermore, hierarchical clustering does not result in clusters that are globally optimal, in that early incorrect choices in linking genes with a branch are not later reversible as the rest of the tree is constructed. So, this method falls into a category known as 'greedy algorithms, 'which provide good answers, but for which finding the most globally optimal set of clusters is computationally intractable.

Despite these disadvantages, hierarchical clustering is a popular technique in surveying microarray expression patterns in an experiment. Self-organising maps. Self-organising maps are similar to hierarchical clustering, in that they also provide a survey of expression patterns within a data set, but the approach is quite different. In other words, each biological sample is considered a separate dimension or axis of this space, and after the axes are defined, genes are plotted using expression levels as coordinates. This is easiest to visualise with three or less microarrays, but extends to a larger number of experiments/dimensions.Nearness can be defined using any of the dissimilarity measures described above, although Euclidean distance is most commonly used.

The process starts with the answer, in that the number of clusters is actually set as an input parameter. A map is set with the centres of each cluster-to-be (known as centroids) arranged in an initial arbitrary configuration, such as a grid. As the method iterates, the centroids move towards randomly chosen genes at a decreasing rate. The method continues until there is no further significant movement of these centroids. The advantages of self-organising maps include easy two-dimensional visualisation of expression patterns and reduced computational requirements compared with methods that require comprehensive pairwise comparisons, such as dendrograms.

However, there are several disadvantages. First, the initial topology of a selforganising map is arbitrary and the movement of the centroids is random, so the final configuration of centroids might not be reproducible. Second, similar to dendrograms, negative associations are not easily found. Third, even after the centroids reach the centres of each cluster, further techniques are needed to delineate the boundaries of each cluster. Finally, genes can belong to only a single cluster at a time.

Relevance networks

Continuing through the set of unsupervised techniques, relevance networks allow networks of features to be built, whether they represent genes, phenotypic or clinical measurements. The technique works by first comparing all features with each other in a pairwise manner, similar to the initial steps of hierarchical clustering. Typically, two genes are compared with each other by plotting all the samples on a scatterplot, using expression levels of the two genes as coordinates. A correlation coefficient is then calculated, although any dissimilarity measure can be used. A threshold value is then chosen, and only those pairs of features with a measure greater than the threshold are kept.

Although the threshold is chosen using permutation analysis, it can actually be used as a dial, increasing and decreasing the number of connections shown. There are several advantages in using relevance networks. First, they allow features of more than one data type to be represented together; for example, if strong enough, a link between systolic blood pressure and expression of a particular gene could be visualised.

Second, features can have a variable number of associations; theoretically, a transcription factor might be associated with more genes than a downstream component. Finally, negative associations can be visualised as well as positive ones. One disadvantage to this method is the degree of complexity seen at lower thresholds, at which many links are found associating many genes in a single network. Completely connected subcomponents of these complex graphs (known as 'cliques') are not easy to find computationally.

Principal-components analysis

Principal-components analysis is more useful as a visualisation technique than as an analytical method. It can be applied to either genes or samples, which are represented as points in multidimensional space, similar to self-organising maps. Principal components are a set of vectors in this space that decreasingly capture the variation seen in the points. The first principal component captures more variation than the second, and so on. The first two or three principal components are used to visualise the gene on screen or on a page. Because each principal component exists in the same multidimensional space, they are linear combinations of the genes.

For example, the greatest variation of biological samples might be described as 3 times the expression level of the first gene, plus -2.1 times the expression level of the second gene, and so on. The principal components are linear combinations that include every gene or sample, and the biological significance of these combinations is not directly intuitive. There are other caveats in using principal components. First, it is important to note whether genes have been centred before analysis; if not, then the first principal component might serve to centre the genes.

Second, it is crucial to note that although principal components might best describe the variation seen in an expression data set, they do not describe how to best separate groups of genes or samples. For example, if microarrays are measured on samples from two conditions, principal components will best describe the variation of those samples, but will not always be the best way to split samples from those two conditions.

Nearest neighbours

Although the nearest-neighbour technique can be used in an unsupervised manner, it is commonly used in a supervised fashion to find genes directly with patterns that best match a designated query pattern. For example, an ideal gene pattern might be one that is highly expressed in one condition and expressed at a low level in another condition.All the genes that have been measured can then be compared to this ideal gene pattern and ranked by their similarity. For example, acute lymphocytic leukaemia was distinguished from acute myelogenous leukaemia using this method.

Although this technique results in genes that might individually split two sets of microarrays, it does not necessarily find the smallest set of genes that most accurately splits the two sets. In other words, a combination of the expression levels of two genes might split two conditions perfectly, but these two genes might not necessarily be the top two genes that are most similar to the idealised pattern. This technique can be modified for specific cases, such as toxicogenomics. Tissue exposed to various compounds known to cause different types of toxicity can be subjected to microarray measurements, as well as measurements from normal tissues. This could make up a training set, in that these data points implicitly make up a model of toxicity. Newer compounds can then be tested on these tissues (the test set), and the 'distance' of these expression patterns from the training set can be calculated, and a decision made as to similarity of mechanism of toxicity.

Support vector machines

Support vector machines address the problem of finding combinations of genes that better split sets of biological samples.Although it is easy to find individual genes that split two sets with reasonable accuracy owing to the large number of genes (also known as features) measured on microarrays, occasionally it is impossible to split sets perfectly using individual genes. The support vector machines technique actually further expands the number of features available by combining genes using mathematical operations (called kernel functions). For example, in addition to using the expression levels of two individual genes A and B to separate two sets of biological samples, the combination features $A—B$, A/B, $(A—B)^2$ and others, can also be generated and used.

To make this clear, it is possible that even if genes A and B individually could not be used to separate the two sets of biological samples, together with the proper kernel function, they might successfully separate the two. Similar to the principal-components analysis method above, consider each biological sample as a point in multidimensional space, in which each dimension is a gene and the coordinate of each point is the expression level of that gene in the sample. Using support vector machines, this high-dimensional space gains even more dimensions, representing the mathematical combinations of genes. The goal for support vector machines is to find a plane in this high-dimensional space that perfectly splits two or more sets of biological samples.Using this technique, the resulting plane has the largest possible margin from samples in the two conditions, therefore avoiding data over-fitting.

It is clear that within this high-dimensional space, it is easier to separate samples from two or more conditions, but one problem is that the separating plane is defined as a function using all the dimensions available. For example, the most accurate plane to separate one disease from another might be $(A—B)^2 < 20$, where A and B are expression levels of genes. Although this might not be the most mathematically accurate way to separate two diseases, the biological significance of such functions is not always intuitive.

TRANSFORMATION OF HYPOTHESES TO BIOINFORMATICS TECHNIQUES

As microarrays now cover the measurable genome, a user does not need to have a set of candidate genes in mind before use. However, this does not mean that a microarray user operates in a hypothesis-free manner.With all the supervised and unsupervised methods that are available for analysis, the challenge is in translating hypotheses into an appropriate bioinformatics technique. In the domains of drug discovery and diagnostic testing, it is easy to find uses for supervised methods. Two examples illustrate this point. First, hypotheses in toxicogenomics, such as 'some genes in the genome influence liver metabolism of a particular compound', could be answered with a technique designed to find those genes for which expression measurements are most significantly different between liver cells with and without the compound.

If many samples are present from each group, a nearest-neighbour approach could be used, in which an idealised gene is created that is expressed at a high level in samples with the compound and at a low level in samples without the compound. This is then used as a query to find genes that are most similar to the idealised pattern. Second, hypotheses in the development of diagnostic tests, such as 'a combination of gene expression measurements accurately distinguishes metastatic from non-metastatic disease', can be answered by searching for such patterns using any of the supervised methods, including support vector machines. Certainly, any specific question can be answered directly in a supervised manner.

The need for unsupervised methods is less intuitive, because these start with a less direct question. For example, questions about the number and type of expression responses in a period of time after application of a compound cannot be found in a supervised manner.Unsupervised techniques, such as hierarchical clustering and self-organising maps, survey all genes and cluster them together on the basis of their expression patterns. A search for the pairs of genes that are most likely to be co-expressed can be accomplished using a technique such as relevance networks. Finally, true genetic regulatory networks might be found using methods such as constructing Bayesian networks. What is common in these three examples is that there is no ideal answer that is being sought, and that it is difficult to ascertain when the method is correct. Nonetheless, unsupervised methods could be instrumental in the early discovery process.

There are a few special cases worth noting.When attempting to find connections between dissimilar items, such as a response to a small molecule (that is, a phenotypic measurement) and expression measurements, two-dimensional hierarchical clustering can be used for each measurement separately, or building a relevance network. Finally,when searching for subtypes of a condition or disease that might influence survival or a disease-free state, unsupervised hierarchical clustering can be paired with Kaplan-Meier survival statistics, as shown by Alizadeh and colleagues.

ISSUES IN MICROARRAY ANALYSIS

After several microarray analyses, it quickly becomes obvious that the rate-limiting step in functionalgenomics experiments is neither the handling of the biological samples nor the actual analysis, but instead the post-analytical work in determining what the results actually mean. First, detailed names and information might not yet be available for genes that have been found to be significant, even though these genes might have been measured on microarrays for years. This complicates the interpretation of results. The official gene name, predicted protein domains or gene-ontology classification might become available as early as tomorrow, or as late as decades from now.

There are further post-analysis challenges. Occasionally, microarray probes are designed against chromosomal regions instead of expressed products, and when these probe sets are positive in an analysis, it is usually not clear which genes are being detected. It is worth finding these probe sets before analysis begins and eliminating them. Occasionally, probe sets are incorrectly designed against the wrong strand or wrong species. Oligonucleotide sequences that were once thought to be unique for a particular gene might not remain unique as more genomic data are collected. Finally, in using spotted cDNA arrays, particularly those for which the probe sequences have not been validated, the findings might be incorrect. Operationally, this means that one is never done analysing a set of microarray data. The infrastructure has to be developed to re-investigate constantly genes and gene information from microarray analyses performed in the past.

REFERENCES

Gavin MacBeath and Stuart L. Schreiber, "Printing Proteins as Microarrays for High-Throughput Function Determination", *Science*, 8 September 2000.

Richard B. Jones, Andrew Gordus, Jordan A. Krall, and Gavin MacBeath, "A quantitative protein interaction network for the ErbB receptors using protein microarrays", *Nature*, 12 January 2006.

8

Genome Annotation

Genome annotation is the process of attaching biological information to sequences. It consists of two main steps:

— identifying elements on the genome, a process called Gene Finding, and

— attaching biological information to these elements.

Automatic annotation tools try to perform all this by computer analysis, as opposed to manual annotation which involves human expertise. Ideally, these approaches co-exist and complement each other in the same annotation pipeline.

The basic level of annotation is using BLAST for finding similarities, and then annotating genomes based on that. However, nowadays more and more additional information is added to the annotation platform. The additional information allows manual annotators to deconvolute discrepancies between genes that are given the same annotation.

For example, the SEED database uses genome context information, similarity scores, experimental data, and integrations of other resources to provide the most accurate genome annotations through their Subsystems approach.

Structural annotation consists in the identification of genomic elements.

— ORFs and their localisation
— gene structure
— coding regions
— location of regulatory motifs

Functional annotation consists in attaching biological information to genomic elements.

— biochemical function
— biological function
— involved regulation and interactions
— expression

CDNA AND PROTEIN SEQUENCES ALIGNMENT

In 1996, Hillier et al. reported sequencing 280,000 human ESTs, thereby increasing the size of GenBank's human EST collection by a factor of six. The ongoing scale-up of EST sequencing created a demand for tools to align these sequences to their genomic sources. While local alignment tools like BLAST can give an approximate answer, determining splice sites accurately in the presence of sequencing error requires algorithms that incorporate a stronger model of the biological processes by which pre-mRNAs are spliced. The simplest approach is to allow "intron gaps" of unbounded length with no gap extension penalty, so long as they begin with GT and end with AG-the dinucleotides that bound 99% of all known introns.

EST_GENOME implemented this simple approach using an algorithm that is optimal. Finding the optimal alignment is computationally demanding, but given the inherent difficulty of the problem, the algorithm used in EST_GENOME is quite efficient. More recent programmes added heuristics that speed up the computation but may not always return the absolute highestscoring alignment. They also added some detail to the scoring system by incorporating an intron-length penalty or a more accurate system of splice-site scoring. There was no clear winner in terms of accuracy.

Lacking a theory about how to set the parameters appropriately for a given degree of similarity between cDNA and genome, the programmes were nearly always used with the default parameters. Perhaps as a result, each system was most accurate for some species and some similarity levels. But the fact that EST_GENOME returns the best alignment according to a simple, clear scoring scheme lends it a unique appeal. Its scoring scheme turned out to be adequate, in the sense that one could not obtain alignments that were significantly more accurate for a broad range of situations, given the sequence quality and computing power available at the time. As an example of the ambiguities that arise in cDNA-togenome alignment, consider a short cDNA segment that can be aligned as the 3′ end of a long exon with mismatches or as a short independent exon without mismatches.

Traditional spliced alignment programmes, such as EST_GENOME will make this decision based on somewhat arbitrary match/mismatch scores and intron penalties that depend on whether the intron begins with GT and ends with AG. A small fraction of introns is known to be bounded by GC-AG (~1%), AT-AC (~0.15%), and even GA-AG (1-3 known cases). Although some AT-AC introns are spliced via the U12 spliceosome, the initial and terminal dinucleotides do not determine whether the U2 or U12 spliceosome is used.

In any case, EST_GENOME does not differentiate among GC-AG, ATAC, or any other intron boundaries except GT-AG. Thus, it will create introns starting with TT under certain circumstances, even though there is no convincing evidence that such introns exist. Compromises like this are necessary when the TT may result from error in sequencing a GT. But when the quality of the genome sequence is high enough, the probability that an intron will start with an apparent TT approaches zero. This illustrates how the best approach to cDNA-to-genome alignment depends on the quality of the sequences involved.

A related gene prediction approach is to align protein sequences or profiles from existing databases to a genome sequence. Because most "protein" sequences in the databases are derived by conceptual translation of cDNAs, and because the alignment algorithms for cDNA and protein sequences are similar, it is tempting to treat cDNA alignment and protein alignment as a single approach to annotation.

However, there is a difference in conception and typical application. Most cDNA alignment programmes are intended primarily for aligning sequences to the genomic locus from which they were transcribed, although these programmes have been used for crossspecies alignments. Protein-oriented alignment programmes, on the other hand, are intended for more distant relationships, such as discovering new members of a known protein family or discovering homologs in a new species.

Philosophically, these are very different approaches. The evidence that a cDNA sequence provides about the exon-intron structure from which it is assembled is much more direct than the evidence that a protein sequence provides about the loci of putative homologs. Cross-locus protein aligners must accept a significant degree of mismatch between the protein to be aligned and the target locus, which can lead to difficulty in distinguishing between functional homologs and nontranscribed pseudogenes. Systems for aligning high-quality cDNA sequences to their genomic sources, on the other hand, can require an almost perfect match, which often helps to distinguish their true loci from nontranscribed pseudogenes.

GeneWise is certainly the most important protein-to-genome alignment programme, since it forms a central part of the highly influential Ensembl gene annotation pipeline. The accuracy of Gene-Wise is determined primarily by the degree of similarity between the protein and the locus to which it is aligned. Thus, several investigators have tried to estimate its accuracy as a function of protein-to-genome similarity. Guigó and colleagues used the P-value of the BLASTP alignment between the protein and the gene locus. The P-value reflects both the length and percent identity of the alignment, but not the fraction of the true gene covered by it.

Using their Semi-Artificial Gene set, which was created by concatenating single gene sequences separated by random "intergenic" sequence, they found that Gene-Wise exceeded the accuracy of Genscan in exact exon prediction only when aligning highly similar proteins ($P < 10^{-50}$). When aligning proteins of moderate similarity ($10^{-50} < P < 10^{-6}$), its overall accuracy was similar to that of Genscan, but GeneWise was more conservative-it missed more of the real exons, but a higher proportion of those it predicted turned out to be exactly right. Instead of using the P-value, Birney et al. measured similarity using percent identity and percent of the target gene covered by the protein. Their most similar category of proteins was 85%-95% identical to the target locus and aligned to within 20 amino acids of the target's start and stop codons.

On a singlegene test set, and excluding terminal exons, they found that GeneWise had 93% exact exon specificity and 75% exact exon sensitivity. Thus, while most of the exons it predicted were correct, it failed to correctly predict one fourth of the real exons. However, two caveats regarding this test are in order. First, it is impossible to know when running GeneWise whether the protein alignment is close enough to the ends of the target to meet the requirements of this accuracy level, since the genes in the target sequence are unknown. If alignments are not selected on length, GeneWise has only 40% exact exon sensitivity using proteins in the 85%-95% identity range. Second, this single gene set is likely impoverished for pseudogenes, which are a major source of false positives for GeneWise.

DE NOVO GENE PREDICTION

De novo gene prediction, in its modern form, also appeared in the mid-1990s. Stormo and Haussler described the first Generalised Hidden Markov model (GHMM) for gene prediction. GHMMs are mathematical models that can be used to define probabilities for all possible exon-intron annotations on a given sequence. An accurate GHMM for gene finding will assign high probabilities to correct annotations and low probabilities to incorrect annotations. GHMMs differ from ordinary HMMs in that the log probabilities, or scores, of exons and introns can depend globally on the entire sequence of the exon or intron.

In ordinary HMMs, the scores of features must be the sums of the scores of individual bases within the feature. For example, the ability to compute scores from the whole feature makes it possible to create general, nonlinear models of the lengths of exons and introns. Kulp et al. were the first to use the term GHMM in the context of gene finding, the first to describe a fully general mathematical framework for all GHMM models, and the first to implement implement and test a GHMM-based computer programme for gene finding.

Burge and Karlin developed Genscan, a GHMM-based gene prediction programme that could predict multiple and partial genes on both strands. This ability made Genscan suitable for annotating ORFs in anonymous DNA sequence such as that produced by sequencing random BAC clones. As the first GHMM suitable for annotating anonymous genome sequence, Genscan defined the state of the art in both technology and accuracy. Eventually, programmes like Genie were enhanced to predict multiple genes on both strands, but Genscan remained one of the most accurate and most widely used programmes for many years.

COMBINING PREDICTION

The genome annotations that are most visible to the public and most widely used are created by combining predictions from all of these methods. In the mid-1990s, the results of cDNA alignments, protein alignments, and de novo predictions were integrated by human experts and were often followed by RT-PCR and sequencing experiments to test the predicted exon-intron structures (RT-PCR is PCR amplification of cDNAs made by reverse transcription from RNA). Automated "pipelines" for integrating evidence, such as Ensembl, OTTO, and the National Centre for Biotechnology Information (NCBI) pipeline were not developed until shortly before the publication of the draft human genome sequence.

Accuracy of Prediction Systems

The accuracy of prediction systems based on aligning cDNA or protein sequence depends on the sequences that are available for alignment as well as the algorithms used to align them. There can be no doubt that both the quantity and quality of expressed sequences have improved dramatically in the last ten years. For example, the human EST database has gone from 415,000 sequences in 1997 to over 6 million in 2005. Several projects, including the Mammalian Gene Collection (MGC) have produced finished sequences from large collections of cDNA clones that appear to contain a complete ORF.

Indeed, the MGC collection now contains at least one cloned transcript from about 13,000 human and 12,000 mouse gene loci. These sequences have been subjected to extremely rigorous quality control so that most produce 100% identical alignments to the reference genome, except for silent discrepancies and known polymorphisms. There have been improvements in alignment algorithms, too. Traditional cDNA-to-genome alignment programmes do not explicitly model the probability of mismatches in the correct alignment as compared with the probability of an additional intron in the correct alignment.

In fact, mismatches in correct alignments are either sequencing errors or differences between the reference genome and the genome from which the cDNA was transcribed. Thus, the probabilities of these events depend on both the sequence quality and the rate of polymorphism or divergence. On the other hand, the probability of an additional intron depends on the frequency of introns in the species at hand. A new generation of cDNA-to-genome alignment programmes models all these things using pair hidden Markov models with parameters estimated from the specific cDNA collection and the genome sequence to be aligned.

For example, such systems can easily model the fact that sequencing errors are much less likely when aligning an MGC cDNA sequence to the finished human genome than when aligning a single-pass EST sequence to the draft dog genome. For 70%-80% of high-quality cDNA sequences, these more precise models will result in the same alignment as a programme like EST_GENOME. In many of the remaining cases, however, they produce better alignments. For example, they are better able to distinguish small exons from sequencing errors. This accuracy improvement is made possible by the availability of very high quality sequences to align and the availability of sufficient computing power to run the pairHMM algorithms in a reasonable amount of time.

Single-genome de novo gene prediction

Perhaps the best indicator of how the accuracy of de novo gene prediction has changed in the last ten years is the change in how accuracy is measured. Shortly before the publication of Genscan, Burset and Guigó published the first comprehensive comparison of vertebrate prediction programmes. Their test set consisted of 570 genomic sequences no longer than 50 Kb, each with a single gene on the positive strand. Burset and Guigó found that the most accurate de novo system of that time, FGENEH, predicted just 61% of the known exons correctly, GeneID got 51%, and all the rest got well below 50% of exons right.

The fraction of ORFs predicted correctly was not reported, presumably because it was near zero. Genscan represented a breakthrough in accuracy that led to a long, slow shift in the evaluation paradigm. When tested on Burset and Guigó's single-gene set, it predicted 78% of the exons correctly, compared with just 61% for the best previous system. Furthermore, Burge and Karlin reported that Genscan predicted 43% of the ORFs in that test correctly. They also presented an analysis of Genscan's predictions on a contiguous sequence of 117 Kb containing multiple experimentally determined gene structures.

Although the number of genes was too small for reliable estimation of accuracy, it is interesting that only one of eight Genscan-predicted ORFs matched the annotation exactly (12%), though most were quite similar. As test sets became more realistic, estimates of Genscan's accuracy at predicting complete human ORFs dropped. Guigó et al. published an evaluation based on simulated human genomic sequence that they created by concatenating singlegene singlegene sequences padded by randomly generated pseudointergenic sequence.

In this new test set, only 2.3% of the nucleotides were protein-coding, much closer to the overall average than the 15% in Burset and Guigó's 1996 set. As expected, they found that prediction accuracy is much lower on contiguous sequences with typical coding density than on single-gene sequences with high coding density. However, they did not report the percentage of ORFs predicted correctly. Korf et al. tested Genscan on 7.6 Mb of mouse genome consisting of 68 contiguous sequences with an average length of 112 Kb. In this test, Genscan predicted only about 15% of annotated ORFs exactly-much lower than the 43% reported for Burset and Guigó's single-gene set. However, even this estimate turned out to be optimistic.

When Genscan was finally evaluated on the entire human genome, it predicted a correct ORF at only 10% of loci containing a known ORF. Currently, gene prediction programmes are used primarily for whole genome annotation. As described above, their accuracy when evaluated on a whole genome is typically much lower than their accuracy when evaluated on isolated genes or artificially concatenated sets of single genes. Even whole chromosomes can be deceptive. For example, human chromosome 22, besides being the smallest autosome, is also unusually gene dense, with smaller than average introns and intergenic regions and above average GC content.

Most gene prediction programmes, including Genscan, tend to perform best on high GC, gene-dense regions. Thus, evaluation on chromosome 22 systematically overestimates the accuracy of most systems. In the current environment, the minimal standard for evaluation of gene prediction programmes must be based on whole genome annotation runs. Some may argue that, since we do not know all the exon-intron structures for the human or any other genome, we cannot know the accuracy of a prediction set for the whole genome. This is true, but it should not be an impediment to evaluating whole genome annotations.

Sensitivity estimates based on the subset of genes whose structures are known should be an unbiased estimate of sensitivity on all genes, to the extent that the sets of known genes and unknown genes do not differ in ways that greatly affect accuracy. While it is possible that unknown genes are radically different from known genes in this way, there is no reason to believe that they are. Specificity will be systematically underestimated when the predictions are compared to known genes rather than to all genes.

Under the same assumption described above, dividing by the fraction of genes that are known corrects the underestimate. The exact value of that correction factor does not matter when comparing the specificities of two programmes—the one with the higher raw estimate will also have the higher corrected estimate. Another approach, which seems to always give qualitatively similar results, is to use only gene predictions that overlap known genes by at least one nucleotide when computing specificity.

Determining gene boundaries is one of the most challenging aspects of ORF prediction-much more so than predicting the boundaries of exons with splices on both sides-and so it is also the area in which the potential for improvement is greatest. Many improvements to gene prediction algorithms have a large effect on accuracy as measured by exact ORF prediction, even though they have little effect on the accuracy of exon prediction. Thus, it is critical to include measures of exact ORF prediction in comparative evaluations of gene prediction programmes.

Statistics on the exact-ORF accuracies of programmes are important, but there is a legitimate argument that the value of these programmes is not in predicting known genes but in predicting novel genes. Thus, the most convincing evaluation of a programme or a set of programmes is the extent to which its novel predictions can be verified experimentally. The trend toward publishing experimental evaluations of prediction sets therefore represents a significant step forward for the field of gene prediction.

Dual- and multi-genome de novo predictors

Historically, Genscan represents the apogee in the arc of

improving accuracy for mammalian gene predictors using a single genomic sequence as their only input. No other system robustly outperformed Genscan on large, contiguous human genomic sequences until the advent of dual-genome de novo systems, which use alignments between two genomes as a rough indicator of which nucleotides are under negative selection and hence are likely to have a function that contributes to fitness. Several such systems required that orthologous stretches of mouse and human DNA be identified in advance, or were tested only on single-gene sets that were preselected to have clearly identifiable orthologs.

However, TWINSCAN and SGP2 could both be run on entire human chromosomes, using alignments generated by simple, robust procedures. Neither programme requires that an ortholog be present in the other genome-they can just as easily exploit similarity to paralogs, or even fragments of several genes from the same family. Flicek et al. reported that TWINSCAN was able to predict 14% of known ORFs in the human genome correctly. Although SGP2 was tested only on chromosome 22, which is unusually gene dense compared with the whole genome, the accuracies of the two systems were similar.

Over the last few years, incremental improvements to TWINSCAN, along with better training and testing sets, have improved its accuracy to the point where it can predict a correct ORF at about 25% of human loci with known ORFs. A new level of accuracy was achieved this year by N-SCAN, a version of TWINSCAN with a new, phylogenetic conservation model that is capable of considering alignments among multiple genomes. N-SCAN is able to predict a correct ORF at about 35% of human loci with known ORFs. It is also notably more accurate than previous systems at the exon level, predicting 85% of known human coding exons correctly, whereas previous systems predicted fewer than 75% correctly.

Furthermore, it is the first programme to accurately predict the boundaries of long introns-it correctly predicts about 50% of introns in the 50- to 100-Kb length range.

Many of the challenges of de novo gene prediction that have been observed over the years remain challenges today. Even the best prediction programmes tend to split and fuse genes, and they have difficulty accurately predicting stop codons and especially start codons. They only predict a single isoform at each locus, even though a large fraction of human genes are alternatively spliced. Yet there has been enormous progress. We have moved from predicting a correct ORF at one tenth of the human loci to predicting a correct ORF at one third. We can now predict long introns, and we can predict spliced 5′ UTRs with reasonable accuracy. Gene predictors are generally more accurate on more compact genomes such as those of Caenorhabditis elegans and *D. melanogaster*, but the last ten years have seen substantial progress there, too.

Combining prediction methods

In the run-up to the initial publications on the human genome, it became clear that manual integration of evidence from various prediction methods would not be fast enough to provide an analysis of the entire genome in a reasonable amount of time. As an alternative, several automated "pipelines" for integrating evidence, such as OTTO, Ensembl, and the NCBI pipeline were developed. OTTO was used primarily by the team at Celera Genomics that developed it. Ensembl annotations have been used to produce the primary gene sets in many of the publications that describe the first analysis of a new vertebrate genome sequence.

The NCBI annotation pipeline is also very influential because its predictions appear in GenBank as RefSeq mRNAs and proteins with "XM" and "XP" accessions, respectively. The NCBI pipeline was originally based on GenomeScan, an enhancement of Genscan that modifies the scores of potential exons depending on whether they have high-scoring alignments to proteins in the databases. More recent versions of the NCBI pipeline use an unpublished method called Gnomon. Although every pipeline works differently, both Ensembl and NCBI rely heavily on aligning protein sequences generated from one gene to the genomic loci of other genes, either within or between species.

Most Ensembl gene predictions are ultimately created by GeneWise, a protein-alignment programme, although Genscan is used to help identify the best proteins to align from other species. Thus, GeneWise has determined the structures of a large fraction of the predicted genes used in the initial analyses of numerous vertebrate genomes. The goal in these analyses was to obtain a conservative set of predicted exons and genes-one containing few false positives. Genscan, the best available de novo gene predictor until 2003, predicts numerous false positive exons, so choosing a protein alignment method such as GeneWise made sense.

Even TWINSCAN, as published in 2003, predicted only 75% of known exons correctly, as compared to 85% for Ensembl. Given the recent progress in de novo gene prediction, it is worth asking whether GeneWise is still more accurate, or even more conservative, than the best de novo predictors. A direct comparison would be most informative, but the data set that Birney et al. used to assess the accuracy of GeneWise has been lost. However, some inferences can be drawn by comparing published results on different test sets.

Gross and Brent reported that N-SCAN, when run on the whole human genome, predicts 85% of all known exons correctly; estimated specificity based on the assumption that the genome contains 200,000 exons is 86%. Considering only internal exons, as in Birney et al., N-SCAN's estimated exon-level specificity rises to 93%. Considering that these numbers are based on whole-genome annotations, whereas the estimated 93% for GeneWise is based on a single-gene set, it is likely that predictions based on protein homology are no longer more "conservative" (i.e., specific) than de novo predictions except when the protein is nearly identical to the target locus.

Furthermore, the sensitivity of de novo methods is much higher. A direct comparison was recently made among integrated annotation pipelines as part of the E-GASP community evaluation. Pipeline predictions were compared to manual annotation by the HAVANA group on the human ENCODE regions. Among entrants were Ensembl and a simple ad hoc pipeline with two stages.

The first stage was to align full-ORF cDNA sequences from the MGC project and the RefSeq collection using a cDNA aligner called Pairagon, which is based on a strong prior model designed for highly accurate sequences. The second stage, applied to regions not covered by cDNA alignments, was to exploit BLAT alignments of spliced human ESTs to guide de novo predictions by N-SCAN. These predictions, made by N-SCAN_EST, are generally consistent with the EST alignments but may extend them with additional exons or link nonoverlapping ESTs into a single transcript.

The results showed that the Pairagon+N-SCAN_EST pipeline was substantially more specific than Ensembl, and equally sensitive, in both the exact exon and exact transcript measures. Ensembl predicted more transcripts per locus than Pairagon+N-SCAN_EST, so it could predict at least one correct transcript at a higher percentage of the loci where it made predictions (gene specificity). Thus, Pairagon+N-SCAN_EST, which uses only human cDNA sequence aligned to its native locus, is at least as accurate as Ensembl, which uses cross-locus and cross-species protein alignment.

However, cross-species protein homology is likely to contribute more to the annotation of species for which fewer cDNA sequences are available. A more recent approach to integrating predictions is to score each potential exon using a weighted combination of evidence from alignment-based predictions and de novo predictions. The weights are derived from estimates of the accuracy of each prediction source. Thus, if several predictors that have proven accurate in the past agree on an exon, it will receive a high score.

In the case of disagreement among predictors the score will generally be lower, but more weight will be given to more accurate predictors. This approach performs well in practice, especially when there are multiple evidence sources with roughly similar accuracy-empirically, it seems that different methods make different errors. Indeed, JIGSAW, a descendent of COMBINER, was slightly more sensitive than the Pairagon+N-SCAN_EST pipeline at the exon level in the EGASP evaluations, although Pairagon+N-SCAN_EST was more accurate in predicting exact ORFs. To achieve the accuracy it did, JIGSAW was run on 13 sources of evidence using genome coordinates provided by the UCSC browser, including Ensembl, RefSeq, Genscan, SGP, TWINSCAN, Human mRNAs, TIGR Gene Index, and UniGene.

Manual annotation has also progressed over the last 10 years. In 2000, the Drosophila community held an annotation "jamboree," in which fly biologists and bioinformaticians gathered at Celera Genomics for two weeks to create an initial annotation of the Drosophila genome. This annotation has since been systematically revised and updated. In 2002, the Sanger Institute held two Human Annotation Workshops. A number of groups involved in human annotation gathered at these meetings and compared their annotations on designated sequences to "define a standard of annotation" and "draw up guidelines to help achieve the standard". These meetings led to the annotation standards that are used by the Sanger Institute's Human and Vertebrate Analysis project.

HAVANA annotators integrate information from alignments of expressed sequences and de novo predictions by Genscan and FGENESH. The HAVANA team has annotated human chromosomes 1, 6, 9, 10, 13, 20, 22, and X; mouse chromosomes 2, 4, 11, and X; and the entire zebrafish genome. Their annotation of the human ENCODE regions appears to have been quite complete, since RT-PCR and sequencing experiments have verified only a small handful of exons not annotated by HAVANA.

AUTOMATED GENOME ANNOTATION

Genome annotation is the assignment of biological functions to as many predicted coding sequences as possible. In addition, a complete annotation of a genome includes information regarding gene location and organisation, the transcripts and products of those genes, and the regulation and control of expression, translation and degradation. Annotation must also in some way present the manner in which genes and their products work together in cellular processes. Annotation is done at two levels, initially at the DNA sequence level and then at the level of predicted protein sequences.

DNA sequence analysis estimates boundaries between coding and non-coding sequence, identifies DNA features associated with gene structures, translates protein-coding genes into protein sequence, and characterises conditions under which different forms of a gene may be expressed. Protein sequence analysis takes annotation a step further. Protein level analysis helps to predict function, to associate cellular processes with genome organisation, and ties back to DNA level analyses by assisting in refining predicted gene structures.

Once the individual coding sequences are discerned, genome annotation constructs systems of genes and gene products by combining knowledge gained from sequence analysis with knowledge from other data sources. For example, pathway databases provide information to associate individual genes predicted in an organism with protein-level networks in the same and other organisms. Even in the absence of pathway information or global similarity to proteins of known function, individual regions of the predicted protein can hint at putative function. Leader domains target proteins for secretion through the cell membrane or transport through a nuclear, mitochondrial, or chloroplast membrane, for instance. Internal protein domains can indicate whether the protein is likely to be involved in signaling. Other protein properties indicate whether the protein is likely to be an enzyme, an intracellular globular protein, or a transmembrane protein. Protein sequence motifs can suggest regions that may be modified or play key roles in regulation and activation. The presence of phosphorylation, glycosylation, cleavage or splice motifs can often support functional assignment and placing a given protein within the context of a larger network or pathway.

Information about genes that are expressed together under different conditions further serves to associate individual gene products with cellular subsystems. Co-expression data helps to confirm that genes do indeed participate in common cell processes. More importantly, co-expression data indicate relationships between genes beyond those that can be deduced from known sequence and pathway information.

Genome annotation is the process of uncovering all of these features at the DNA and protein sequence levels and integrating them into a systematic view of the organism's molecular machinery. The very first challenge in annotating a genome is separating the information coding portions of a genomic sequence from the non-coding sequences. In all organisms, coding regions are differentiated from neighboring non-coding regions by specific features. Detecting these features is essential to transforming sequence data into knowledge of a genome. Once a gene is identified or predicted, the next step is to identify putative functions, possible homologs in other organisms and within the genome, and to postulate its role in the life of the organism. These tasks require the use of multiple bioinformatics tools. Gene finding tools assist in identifying genes, and the analysis of gene products via homology-bases searches and other methods allows identification of functions and roles in the larger cellular context.

GENOME ANNOTATION PROCESS

Genome sequence annotation is an iterative process. It requires finding putative coding regions, confirming that these regions are truly coding regions, identifying what each encodes, and using the available evidence to refine the coding regions. Each of these steps involves collection of data from specific software tools that identify genomic or protein features. All of the steps must be controlled by a meta-level process that decides what tools to run, what input to give to each tool, and how to synthesise the output from all the tools into coherent reports about proteins and protein systems.

There are four essential data collection steps: parsing genes from the genome, homology searches, organisation and evaluation of the data, and the generation of an annotation (automatic or human-directed) based on the evidence collected. The final step remains human-intensive, but no thorough evaluation of a genome could (or should) be accomplished in an entirely automated fashion.

Pre-processing genomic sequence data is the first step in annotation of a eukaryotic genome. First, the genomic sequence must be split into overlapping contiguous sequences (contigs) whose length is acceptable as input to genomic sequence analysis tools. Next, repeat regions are identified. Masked and unmasked contigs are sent through genefinding tools to generate exons and extract spliced coding sequences.

Each coding sequence is compared with sequences in the public databases at both the DNA and protein level. Expressed sequence tags, full-length and partial cDNA sequences, and known protein sequences are compared to the predicted DNA sequence, while protein sequences translated from each coding sequence are searched for patterns that indicate functional and structural features. The results of these homology-based searches are used in developing a comprehensive picture of the putative function for the coding region.

Evidence collected from these methods and other bioinformatics tools provides the means for generating a logical model of each gene's structure and a possible functional assignment for the protein encoded by that gene. The gene model includes predicted promoters, transcription factor binding sites, terminator sequences, 5′ and 3′ UTR, poly-adenylation sites, intron boundaries, and, when possible, alternative splicing of exons. At the protein level, evidence is used in predicting putative binding, post-translational modification sites, and, when possible, subcellular localisation.

These individual evaluations and function assignments are then used in more global evaluations of the genome annotation. At this stage, one can assess the completeness of known metabolic or signalling pathways for instance. If a given pathway is incomplete, it might indicate that a region of the genome was not analysed, or that some evidence was overlooked. It may also hint at unusual biology or alternative pathways in that particular organism. These meta-level analyses build upon the initial annotation and develop a genome-scale knowledge base available for others to evaluate and use in their own annotation efforts.

Coding and Non-coding Sequence

The initial annotation step is to find likely coding regions with gene-finding tools trained for the organism. Any series of non-stop codons can be translated computationally into an amino acid sequence. If such a region, called an open reading frame or ORF, begins a gene, it starts with a start codon. The first translated codon in a eukaryotic gene encodes methionine (ATG); in prokaryotic genes, the first codon is usually ATG and sometimes TTG or GTG. Gene finding tools use a variety of algorithms and computational techniques to assess the overall frequency of each possible codon in known genes of a particular genome and downgrade or upgrade translatable regions according to codon usage.

For eukaryotic genomes, other information such as splice site patterns, average length of introns, and presence of poly-adenylation sites help to discriminate coding from non-coding regions. In prokaryotes, ORFs can be easily defined syntactically; they are the regions of the genome that contain a start codon, a series of translatable codons and a stop codon. In addition, one can specify certain hallmarks of "true" coding regions. For instance, most ORFs contain the highly conserved Shine-Delgarno sequences for improved translation of the transcript. In addition, the percentage of rare codons tends to be relatively low in genuine coding regions. These and other criteria can be used to produce a fairly robust ORF-finding tool.

As with prokaryotes, eukaryotic genomes contain signals that delineate genuine coding regions from the non-coding regions of the genome. However, the simplistic definitions of an ORF that are satisfactory in prokaryotes almost always fail to capture true coding regions for eukaryotic genes because of the presence of introns. The presence of introns makes gene prediction computationally much more complicated because the actual coding sequence is split across multiple open reading frames.

There are three well-known gene prediction tools, GENSCAN from Burge and Karlin, HMM gene from Anderson et al, FGENES and FGENESH from the Sanger Center's Victor Solovyev. Included is a BLASTx search (translation in all six frames of the nucleotide query and search against the non-redundant protein database) of a 100kbp test region of eukaryotic genomic DNA. As becomes obvious from this simple comparison, while all the tools are within a "ballpark figure" of the number of likely coding regions in this test sequence, the final numbers do vary somewhat from tool to tool.

GENSCAN predicted the fewest exons, although FGENESH was not far behind. However, GENSCAN produced a more parsimonious set of gene structure predictions. Indeed, as the documentation for FGENESH notes, this tool is best at finding the individual exons, but does less well at stringing any given set of exons into a coherent gene structure. HMM gene did the most poorly of the three in this very simplistic comparison, although a brief perusal of the actual predictions suggests that it has a bias toward finding several very small exons (between 50 and 150bp) around what the other tools identify as a single, larger initiating exon. Regardless of such variations, what this simple comparison does demonstrate is that gene prediction in eukaryotes is still as much an art as it is a science.

Some regions of the gene structure can be more easily confirmed than others. Initial and terminal exons and intron-exon boundaries may be confirmed by matches with sequences known to be expressed in the organism.

Alignments with known 3′ and 5′ ends of expressed sequence tags (ESTs) from cDNA clones anchor ends of genes and indicate alternative ends of transcription for the same gene. The use of ESTs in gene refinement is discussed further in the sections that follow. Other features within the DNA sequence help to signal which coding regions are contained within a single transcript. For example, in eukaryotic genomes, poly-adenylation sites after a terminal exon and transcription initiation sites upstream of the first coding exon indicate transcription boundaries. In prokaryotic genomes, a series of coding regions may be transcribed together as an operon and then each translated as independent amino acid sequences after transcription.

Non-coding Regions of the Genome

Even if a gene's coding region is correctly predicted, additional information may be required to predict the actual transcript and protein product. This is especially true for higher eukaryotic genomes, where much more post-transcriptional and translational modification occurs in the process of generating a functional protein from the genomic DNA sequence of the gene. A short list of such modifications includes alternative splicing at the DNA and RNA levels, RNA editing, post-translational modifications to the proteins and many other variations on these themes. All these manipulations of a coding sequence must be included in the computational analysis of a eukaryotic gene if it is to be of any use to the biologist. The non-coding regions of eukaryotic genomes are also rich with variation. Non-coding regions in eukaryotes play significant roles in the biology of the organism, containing regulatory regions with promoters and transcription factor binding sites. Less obvious and harder to identify computationally are non-coding regions with structural features that may be valuable for overall regulation of a genomic sequence, such as histone binding sites and larger structures including centromeres and telomeres.

Protein Level Analysis of Coding Sequences

Once coding sequences are identified and extracted from the genome, evidence for each coding sequence (CDS) is collected from a variety of sequence homology and motif search tools. The output from each tool is integrated into a generic relational format. The most common tools currently employed for homology searches include BLAST, FASTA, and BLAIZE searches against the public nucleotide and non-redundant protein databases. Searches of the dbEST expressed sequence tag (EST) database provide valuable information to confirm the transcription and splicing of predicted open reading frames even in the absence of protein or direct nucleotide homologies. EST searches can be done at the DNA level with BLASTn or FASTA or at the protein level via tBLASTn, tBLASTx or FASTAx.

Protein motif identification can be done with searches against the PROSITE, pFAM and BLOCKS databases with PROSEARCH and the pFAM and BLOCKS search software. Finally, a survey of RNA genes can be done using tRNAscan-SE to identify tRNA genes and BLASTn against the rRNA subset of Genbank to identify rRNA genes. If protein or DNA level sequence homologies are sufficiently strong and informative, their functional information can be transferred to the genomic query sequence as a putative functional annotation. Alternatively, function can be postulated based on the collective evidence of other non-homology based analyses. Tools that inspect the protein for structural and subcellular localisation features help to determine the cellular location and role of the protein.

If a portion of the protein sequence matches a domain of a protein of known three-dimensional (3D) structure, a 3D homology model can be constructed for the protein. A 3D model serves to confirm or overrule function assignments and to gain further insight into how the protein may bind with other molecules. Structure prediction from homology is complemented by attempts at ab initio structure prediction which is done without prior knowledge of homologous protein structures.

Approximately 4,200 genes annotated in *E. coli*, nearly 40% lack any functional annotation. In organisms such as the anciently diverged classes of eukaryotes, the number of ORFs with effective functional designations is even lower. Therefore, homology based searching will not provide as comprehensive a view of the genome as would be desired. Yet, even in cases where strict homology-based annotation is impossible, accumulated evidence from several tools can sometimes provide as useful an annotation. The key to developing supportive evidence of the "authenticity" of a predicted coding region is in identifying the hallmarks of a true eukaryotic gene. While these sequence motifs and structures are not as highly conserved from eukaryote to eukaryote, some are conducive to computational analysis and identification. The primary challenge for coding regions with relatively poor homology-based evidence is confirming the validity of the prediction.

Gene Refinement

The most difficult genome annotation task is to determine exact boundaries when separating coding sequence from non-coding. In most eukaryotic organisms, boundary predictions are complicated by the fact that multiple coding regions often comprise the complete gene encoding a protein, and different combinations occur under different conditions. In prokaryotic organisms and some microbial eukaryotes, each open reading frame generally corresponds to one coding region, but the start of the actual coding region can be ambiguous. If gene boundary signals are weak, ambiguous, or not detectable, distinct genes may be merged, exons can be assigned to the wrong gene, and single genes can be split into multiple genes. As each predicted gene product is compared to protein sequences, motifs and 3D structures, the alignments help to refine boundaries between coding and non-coding regions. Of particular value in these evaluations are alignments to ESTs.

REFERENCES

Arun Jagota. *Microarray Data Analysis and Visualization*. Bioinformatics by the Bay Press. 2001

Bishop, J.E. and Waldholz, M. *Genome: The Story of Our Astonishing Attempt to Map All the Genes in the Human Body*. Touchstone Press, New York. 1990.

Isaac S. Kohane, Alvin Kho, Atul J. Butte. *Microarrays for an Integrative Genomics (Computational Molecular Biology*. MIT Press. 2002.

Sensen C., *Essentials of Genomics and Bioinformatics*, John Wiley & Sons. 2002.

9

Challenges and Opportunities of Bioinformatics

We are currently witnessing a technological revolution. With the increase of sequencing projects, bioinformatics continues to make considerable progress in biology by providing scientists with access to the genomic information. This progress is especially contributed by the Human Genome Project. The information obtained with the help of Bioinformatics tools furthers our understanding of various genetic and other diseases and helps identify new drug targets. With technological developments of the Internet, scientists are now able to freely access volumes of such biological information, which enables the advancement of scientific discoveries in biomedicine.

In spite of being young, the science of Bioinformatics exhibits tremendous potential for playing a major role in the future development of science and technology. This is evident from the fact that modern biology and related sciences are increasingly becoming dependent on this new technology. It is expected that Bioinformatics will especially contribute in the future as the leading edge in biomedicine to pharmaceutical companies by expediently yielding a greater quantity of lead drugs for therapy.

DATAMINING VS MODELLING

One of the main activities underlying bioinformatics is

datamining, namely the use of computers to seek automatically for patterns, structures and rules in large databases. As such, datamining can be understood as an extension of the area of pattern recognition, where one studies methods for identifying patterns in not necessarily so large databases. The essential problem of pattern recognition can be easily understood from the three simple diagrams in Figure 1.

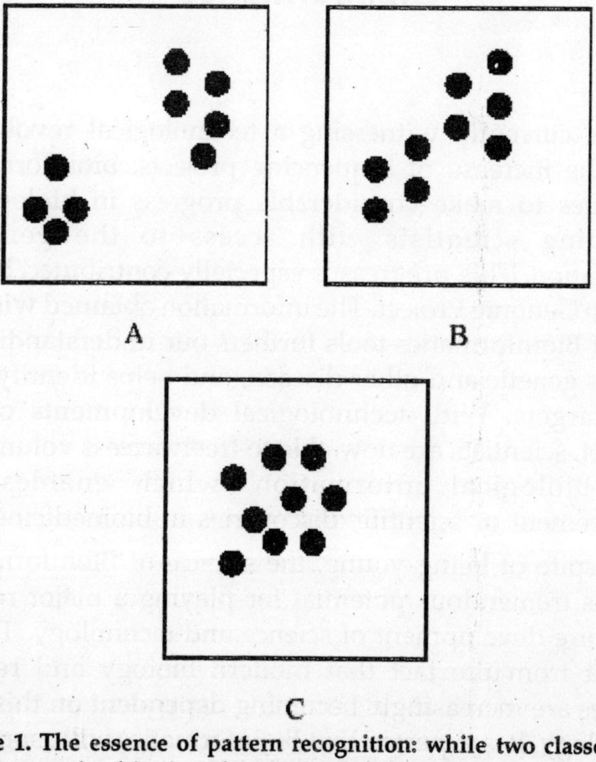

Figure 1. The essence of pattern recognition: while two classes are clearly perceived in A and a single class is seen in C, there is no means for a human to tell if B corresponds to one or two classes

While two groups are immediately perceived in Figure 1A and a single group is seen in Figure 1C, it is rather difficult to conclude whether the dots in Figure 1B belong to one or two groups (you will very likely get different answers from different people). As the science of trying to find an answer to such an issue, pattern recognition stands out as one of the most difficult and subjective scientific activities. Although elegant mathematic and statistical concepts and methods have been developed and applied to cope with this important problem, the subjective nature in which problems are usually formulated—combined with a number of other issues, such as statistical sampling and the decision to use normalised or dimensional data, often conspire to undermine the whole enterprise, or at the very least make it a highly subjective.

The important point to be borne in mind is that, unless some objective mathematical merit function is supplied, the solution of the difficult case in Figure 1B will ultimately depend on human judgment (e.g., Gestalt), which is inherently subjective. This point is especially important because very often the human inspection of the obtained measurement spaces, achieved through scientific visualisation, completely defines the conclusions regarding the structure of the clusters in the data. As a successor to pattern recognition, datamining often suffers from the same inherent problems of subjectivity, now amplified by the large size of the database and the more automated nature of the methods adopted for the mining.

Re-integrating Nature

While biology is a single connected whole, scientific approaches to understanding biological phenomena have often relied on reductionism. Such an approach is characterised by the focus of interest on some particular part of the phenomenon of interest, which is isolated from the rest of nature as thoroughly as possible. Such a divisive and isolationist approach is interesting and effective because, by limiting the effect of other elements on the problem of interest, it enhances our chances of understanding the problem and obtaining a suitable model. Although the reductionist approach applied to biological systems has provided a wealth of findings and knowledge about nature, it is reaching the rock bottom of specialisation.

In other words, to continue learning about a specific biological system it is necessary to look around and make connections with other systems and structures. After all, life is a direct consequence of a long, dynamic, evolutionary process, involving a very wide range of spatial and temporal scales. How can we understand the beak of a given species of hummingbird without considering the flowers from which that small creature feeds? And, to understand the flowers, we need to go further and consider the environment where they developed. Ultimately, it is impossible to isolate life from the rest of our universe in the medium and long term.

Complex Networks

The new area of complex networks, which can be understood as the integration of graph theory and statistical physics, is poised to become one of the most interesting research areas in this century. The origin of the area of complex networks can be traced back to the pioneering works of Erdös and Rényi on random networks. Such discrete structures can be obtained as follows: given N initially isolated nodes, links are added at random between any two nodes.

Interestingly, as the number n of edges increases, the nodes become more and more connected, and trees and cycles start to appear. At a critical edge density, the network undergoes a major topological transformation, in the sense that a giant cluster appears. Such an abrupt change of the properties of the random network is a nice example of the concept of phase transition, more specifically a percolation.

Although Erdös and Rényi accomplished a very comprehensive mathematical study of random networks, such structures were unfortunately found not to be good models of natural structures or phenomena. Later, developments by Watts and Strogatz and other scientists from various areas targeted the interesting type of networks characterised by a relatively small average number of edges between any two nodes, which were duly called small world networks.

While such networks were successfully used to model social relationships, the small world property turned out to be rather general, being also characteristic of random and many other types of networks. Recent developments by Barabási and other scientists led to the interesting network model known as scale-free networks. Such structures are characterised by the fact that they do not have a typical node degree, which is formally defined as the number of edges attached to a node. Indeed, dilogue plots of the distribution of node degree are found to converge to straight lines, without any typical trend. Such a structural property has the important implication that nodes with a high degree—the so-called hubs—become more likely to appear. Hubs are all important because they dominate the connectivity of the network. For instance, an attack on a few hubs will quickly dismantle a scale-free network (and also many other types of networks).

Complex networks have been applied to several important problems in bioinformatics, with encouraging success. For instance, it has been shown that protein-protein interaction networks typically follow a scale-free topology, and that essential proteins tend to correspond to hubs. Promising results have also been obtained for modelling and characterising gene activation networks. Although a great part of the initial attention given to complex networks was motivated by the fact that several important systems, such as the Internet, tended to follow scale-free topologies, interest has recently been focused on a series of complementary investigations, such as the identification of communities in networks (i.e., groups of more intensely connected nodes), the study of the dynamics of processes defined over the network topology (e.g., systems of differential equations where each node corresponds to a variable and the edges represent the coupling between such variables), and the extension to all types of networks, including those in which the nodes are allowed to move in the spatial domain.

Special attention has also been drawn to generalising the concepts and measurements of complex networks, including the possibility to define node degree and clustering coefficient to subsets of the networks. Such trends, allied to the inherent potential of graphs for representing virtually any discrete structure (including trees, vectors, lists, queues, and so on), as well as the possibility to run the quite varied dynamics over such structures, makes complex networks a primary candidate for integrating several scientific areas, and perhaps providing a basic representational framework leading to the unification of science.

Neurobioinformatics

While there is more to biology than genomes, the remarkable scientific-technological advances in this area along the current decade have implied that most biological investigations are now expected to incorporate the genetics component, i.e., to be viewed also from the gene perspective. Indeed, genomes contain the set of genes, which can, under specific circumstances, be activated during the life of an individual. As such, genomes are essential in defining the potential for gene expression. However, the understanding of gene activation dynamics requires the consideration, not only of the genome and the environment, both external and internal to the individual, but also of the representation and modelling of the intricate gene regulations coded into the molecular biology of the individuals.

In this sense, gene expression networks become remarkably similar to neuronal networks, which are also characterised by facilitation and inhibition between the variables (the "genes"). At the same time, a better understanding of gene activation and animal development is paving the way towards a more comprehensive understanding of the largest and most sophisticated neuronal network: the human brain. Indeed, a substantial portion of current research in neuroscience targets or at least involves the associated molecular and genetic aspects. Needless to say, many such investigations now involve the extensive use of informatics, hence the name neuroinformatics.

This interplay between the advances in bioinformatics and neuroinformatics defines an interesting positive feedback between these two areas, which should, with time, lead to an exciting synergy between bioinformatics and neuroinformatics, which we shall call neurobioinformatics. As it turns out, such integration is most welcome, as it is badly needed in order to enhance the nearly saturated human intellectual abilities. While in the short term our intelligence is likely to be augmented by informatics—e.g., access to databases and artificial intelligence software through cameras and projectors installed in glasses, or even through direct bionic implants in the nervous system, advances in the long term may include the genetic redesign and improvement of the architecture of our central nervous system. There will be no more excuses for forgetting the birthday of your neighbour.

Human Resources Formation

The first important point to note about human resources is that this is, by far, the most important element in science.

Indeed, while every piece of equipment is ultimately destined to become obsolete, well-trained and intended human beings are destined to, like wine, become better and better with time. At the same time, a well-formed scientist or technician acts as a source of training herself, therefore establishing a multiplicative system of knowledge transmission. However, given the myriad of areas involved in bioinformatics, a second important issue arises, namely: how to train bioinformatics professionals? Should a mathematician become a biologist or vice-versa?

While we wait for the fulfillment of neurobioinformatics, which may ultimately give us the required almost unlimited intellectual ability, the most effective way to conduct multidisciplinary research is through integrated teams of scientists and technicians who are trained in complementary areas but who also share a basic language allowing them to communicate the problems and work together on the interpretation of the results. Still, the question of how to acquire such a common knowledge remains. Of particular importance is how to provide such knowledge to people (grads or undergrads) with predominant basic training in the biological and exact sciences?

Of course, we need to teach some math to biologists and some biology to exact scientists. But to what extent should we do this? While there is no doubt that these two branches of science are equally difficult (or easy), exact sciences have a more vertical nature, in the sense that the learning of one specific subject (e.g., complex variable calculus) requires a previous acquaintance with a long chain of preliminary subjects (e.g., linear algebra, real calculus and complex variables). Biological sciences, in turn, are characterised by a more horizontal nature, implying a wide range of knowledge (e.g., organic chemistry). Therefore, it appears to be the case that familiarisation of biologists with exact sciences should start as soon as possible along the training programme. On the other hand, it is interesting to expose exact scientists to a broad perspective of the biological world.

After all, the application of informatics in biology is likely to quickly extend to all biological areas. In this sense, the perspectives for employment as a researcher or practitioner will always be enhanced, not only by an in-depth knowledge of some areas of expertise, but also by the flexibility and generality of knowledge in complementary areas.

Impact of Research on Industrial Perspectives

The vast prospects for profits stemming from bioinformatics have implied from its beginnings that a good portion of the research aimed at short term impact (i.e., issues currently leading to strong commercial implications with great potential for short term investment return) would be performed by research institutes associated with and/or maintained by major companies, to the point that it becomes a difficult challenge to compete (or complement) research done by such consortia.

A direct consequence of such a state of affairs is that the sponsoring companies will register many of the findings as patents. So, while academia and governmental research institutions should keep involved with such developments, their chances of success are more likely in longer-term research projects. An interesting alternative approach for the smallersized research institutes and companies is to try to identify issues that are currently of little interest but which may, in the medium term, transform themselves into important trends.

In other words, great creativity is required from everybody. The public availability of data and results, accessible via internet, should continue to play an essential role, not only for catalysing discoveries, but also for facilitating research for both small research teams and developing countries.

Issues in Science and Humanity

Although corresponding to the most important issues in science and humanity, relatively little attention has been focused on ethics and moral issues. Currently, one of the major approaches to ethics and moral states that scientists should concentrate on the development of scientific results, leaving its use for others (typically politicians) to decide. However, would not those issues, subjective as they may be, constitute themselves a primary object of interest of science? We should not forget that, after all, science is from humans to humans, or to what humans may in the end become.

FUTURE TASKS IN BIOINFORMATICS

Designers and Evaluators investigate the biological nature and syntactic structure of questions asked and the tasks commonly performed in the field of bioinformatics. From this, a set of evaluation principles for bioinformatics query systems were derived. Such principles will be useful to both designers and evaluators of general bioinformatics querying tools—when designing query based systems it is essential to know what range of queries to offer and the mechanisms needed for their support. These evaluation principles allow both of these aspects to be assessed in bioinformatics query applications.

A questionnaire survey of biologists, in academia and industry, was undertaken to gain a representative set of queries and tasks. A classification of the biological content of these tasks gave the scope of biological topics or semantics that should be covered by a general query system. The syntactic or mechanical nature of the biological tasks can be derived from these data. This reveals the common processes or components in bioinformatics tasks, showing what happens in these tasks but not how they are performed. These common components describe, at a high-level, all bioinformatics tasks in terms of filters of, and transformations on, data. The two aspects of this analysis afford a means of assessing bioinformatics tools that offer a general query service over the many sources and tools.

A general questionnaire was developed to assess the bioinformatics knowledge and usage of the community. A complete set of answers collected may be found at the questionnaire web site. The relevant questions were:

1. What tasks do you most commonly perform?
2. What tasks do you commonly perform, that should be easy, but you feel are too difficult?
3. What questions do you commonly ask of information sources and analysis tools?
4. What questions would you like to be able to ask, given that appropriate sources and tools existed, that may not currently exist?

Questions one and three were placed at distant parts of the questionnaire, within different contexts, in order to obtain as wide as possible a collection of tasks. Question two was used to find which tasks could be either improved or needed to be addressed. Question four was asked to extend the data collected from question one. The questionnaire was constructed according to standard guidelines and distributed to both academic and industry based biologists, in equal proportions.

All were working biologists, rather than bioinformatics specialists, that is, bioinformatics was not their only pursuit. The responses from both the questionnaire and interviews were collected and duplicates due to the repeated questions removed. This gave a total of 315 biological tasks. In order to counter the lack of detail from responses, a small number of person to person interviews were undertaken. Of the 35 respondents, five were selected, simply by availability, for full interview. The questions gathered by this means are marked with a 'i' next to the question number in the summary results.

The responses were placed into broad categories and a set of task summaries were derived, that represented the range of tasks presented in the responses. Table 1 shows the broad classification of responses and the number of occurrences in each category. The classification was based upon a mix of biological task and the bioinformatics used to perform the task. So, multiple sequence alignment is seen as a task, irrespective of the biological purpose to which it is put (pattern finding, for instance). Similarly, searching for non-coding regions of DNA was set as a biological task, separate from similarity searching etc., purely because it had a large number of responses and thus merited a classification of its own.

Three tasks accounted for 54% of reported tasks. These were similarity search, multiple pattern and functional motif search and sequence retrieval. These cover the basic tasks of obtaining a sequence, finding what exists that is similar and what patterns are present that might indicate sequence function. These tasks are subsumed into 'what is the function of my sequence'. Multiple sequence alignment and DNA analysis (Gene finding, restriction mapping, etc.) forms the next largest grouping. These tasks, particularly multiple sequence alignment, are basic techniques for analysing and manipulating sequences.

The responses from the questionnaire primarily reported what appear to be single task queries. In addition, little detail was reported on how sophisticated the users were in their use of the sources. For example, the overwhelming majority of the responses in the similarity search simply stated 'similarity search' or described which type of sequence was searched. There were reports of restrictions placed on the search, such as from a certain species or other sequence collection, or at a certain level of identity. Other tasks were similarly undetailed, but the task summaries show there were a variety of specialisations requested by users.

Other observations from respondents included requests for more sophisticated means to view results and frustration in interoperating between databases. This seamless moving of data between information sources and analysis tools is of great importance when building more complex queries. Though often stated as single tasks, many of the tasks described do, in fact, involve more than one step (collecting sequences for an alignment, before proceeding with primer design or phylogenetic analysis). The request to view results, especially those intermediate results of multi-source tasks, and have flexible, meaningful results display is obviously important.

The only major gap observed in the biology covered by the responses was in the usage of genomic data. This specific area was covered in questions 18 - 20 of the questionnaire. Of those that answered the questions about use and expected use of genome information, (10%) felt genome data had not had a great impact, but expected it to do so in the future. The most frequent expected use was for searching for sequence homologues and for identification of cloned fragments. Only a few, industry based, biologists expected to use genomic data for its own sake: For inter-genome comparison; gene cluster analysis and genome evolution. Within these tasks are those that respondents wanted to be able to ask, given that the relevant technology and data were available.

An interesting aspect of some of the replies was that the tasks requested are already possible. For example, predicting transmembrane regions. Other information, such as finding if a gene is essential, or exploring gene expression data is a case of having publically available, computationally accessible data. This indicates a lack of knowledge about what can currently be achieved by bioinformatics. Many of the requests for future tasks were to group together already possible tasks to run in parallel. A typical example was: for a given DNA sample perform the following analyses: Perform similarity search; find presence and location of exons; identify repeats; identify GpC islands, and translation of sequence at specific stop/start conditions. Others asked for multiple current tasks to be performed in series. For example:

"Identify homologues of a sequence; of these pull out either n closest or sequences specified; align them, giving various output options; put into various phylogenetic/dodistic packages. The results from these analyses could go forward into other analyses, depending on results." This sort of task emphasises the need to be able to interoperate between collecting sequences, analysing those sequences and viewing intermediate results to determine subsequent routes through an analysis.

The biology or semantics of tasks tells us 'what is wanted' and the corresponding syntax tells us 'how to do it'. This syntax should be able to describe the structure or mechanism of any query. Such a description has two uses: first, it describes what components a query system should have to fully satisfy biologists requirements and second, it can be used to describe a benchmark set of query templates or patterns that occur in biological queries. Information from the questionnaire and interviews yields information on the requirements for such a syntax. The syntax must:

— Have the components appropriate to answer all the queries, together with specialisations;
— allow automatic interoperation between parts of a query without necessitating user intervention;
— allow, but not demand, user intervention for reviewing intermediate results;
— perform intermediate format transformation etc. that are a necessary part of interoperation.

Such task components will only describe tasks at a high-level. They do not describe how a particular component works, for example a functional motif search, but simply state that there is a generic component that performs this sort of bioinformatics task. All task components can be regarded as processes that take a collection of objects and return a collection of objects. A collection is either a set, list or bag of objects.

The behaviour of the process may be modified by the setting of parameters. These processes can interoperate, i.e., they can be joined together to perform larger, more complex tasks. So, a collection produced by one process can, given the appropriate semantics, act as an input collection to another process. These collections of data objects are either transformed or filtered into collections of either newly transformed collections of data objects or restricted collections of the same data objects respectively. These components may also contain mechanisms for describing criteria such as 'at least n results' and boolean operations.

Many newly industrialising countries (NICs) start with some availability of basic factors like raw materials or people. Building these is also difficult in a situation where demand conditions are weak, and related and supporting industries are facing similar problems to the industry under focus. Local firms have typically been protected and thereby sheltered from competition or, in more open markets such as in Latin America, swamped by multinationals. There are thus significant barriers in the way of becoming globally competitive in an industry. In spite of this, there have been several examples of successful "latecomer industrialisation."

Japanese success in the automobile industry, Korea in the consumer electronics industry, Taiwan in the semiconductor industry, and more recently, India in the software industry are examples of countries (and their local firms) creating a national competitive advantage. What is interesting to note in these cases is that none of them were based on large investments in research and development or what may be called science-based innovation.

Korea first targeted relatively mature industries such as the chemical industry and then slowly moved into more sophisticated industries like automobiles, shipbuilding, electronics and telecommunication. Korea's success in such diverse industries has been explained in many different ways. One school of thought argues that Korea distorted prices and thereby created incentives for industrialisation in new areas. Another school of thought believes that the Korean government has played a key role on multiple dimensions—by supporting the chaebol with cheap funds; by protecting the domestic market from foreign competition; by supporting the creation of domestic technological capability, etc.

Cutting across these different perspectives, what emerges is a strategic partnership between the state and industry in which "national champions" were allowed to take on high levels of debt and provided protection but were in return expected to develop a strong export orientation and, later, focus on technology development. Local firms created an absorptive capacity within themselves, and created technological capabilities through leveraging partnerships with a wide variety of international players. More importantly, the firms did not remain content with absorbing what they learnt from their partners but went on to improve upon these technologies and become global players in their own right.

The Taiwanese story is similar except that large national champions are replaced by networks of smaller companies and a greater role is seen for public sector technology organisations in absorbing and diffusing technologies. Common to both Korea and Taiwan is the creation of an institutional capacity to promote the absorption and diffusion of diverse capabilities.

Singapore has been able to attract top multinationals to set up manufacturing facilities and hopes to use its Economic Development Board and state-sponsored venture capital to diffuse capabilities to local firms. Mathews and Cho propose that industries having characteristics of rapid turnover of products and high levels of competition, predictability of technology trajectories, availability of product and process technologies, and the availability of leverage trade-offs are likely to be good candidates for application of the leverage-based approach. While the lack of resources could be an impediment to industrialisation and market access could be another barrier, latecomers can have certain advantages as well. For one, they are not locked into existing technologies or different forms of organisational and institutional inertia. Support from the state and the creation of new institutions can help newcomers leapfrog existing players as demonstrated by France and Russia in the nineteenth century.

Changes in global markets for goods and services also provide opportunities for companies from NICs. As competition has become more intense, there has been strong pressure on companies to focus on those activities that they are best at doing and to outsource all other activities. For example, the shoe company Nike focuses on shoe design and the entire chain of marketing activities in which it believes it can create maximum value. It gets all its manufacturing done by companies in low wage cost economies and only supervises the quality of production. In hyper-competitive conditions, no company can afford to be uncompetitive at any stage of value chain.

Firms from NICs have seized this opportunity to become part of the global value chain of buyers. Mathews seeks to identify the process of evolution of latecomer multinational enterprises—companies that have in a short span of time established themselves as significant global players in spite of the lack of resources faced by them in their countries of origin. Their ability to capitalise on the opportunities offered by globalisation has been enhanced by their willingness to experiment with a number of strategic and organisational innovations.

Bioinformatics has emerged in recent years as an industry in its own right. Bioinformatics companies provide the computational tools and databases that made genomic research possible. A 2000 estimate of the bioinformatics industry put its size at $300 million generated by 50 companies, evenly divided between data suppliers and IT/tool providers. This is expected to rise to about $1.5-2.0 billion by 2005, in addition to about $2 billion of internal spending on IT infrastructure by biotechnology and pharmaceutical companies. In addition to companies that focus on bioinformatics, IT and IT service companies like Sun Microsystems, IBM and Agilent Technologies have entered the arena of research and bioinformatics solution providers, either on their own or in alliance with genomics companies.

REFERENCES

Charles Staben, *Bioinformatics: A Primer*, Jones & Bartlett Pub. 2002.

Des Higgins (Editor), Willie Taylor (Editor), *Bioinformatics: Sequence, Structure, and Databanks: A Practical Approach*, Oxford University Press. 2000.

Luc T. Wille (Editor), *New Directions in Statistical Physics: Econophysics, Bioinformatics and Pattern Recognition*, Springer Verlag, 2002.

Bibliography

Arun Jagota. *Microarray Data Analysis and Visualization*. Bioinformatics by the Bay Press. 2001

Bishop M. J., (Editor) *Genetics Databases*. Academic Press. 1999.

Bishop, J.E. and Waldholz, M. *Genome: The Story of Our Astonishing Attempt to Map All the Genes in the Human Body*. Touchstone Press, New York. 1990.

Bowie JU, Luthy R, Eisenberg D., "A method to identify protein sequences that fold into a known three-dimensional structure", *Science*, 1991.

Charles R. Cantor, Cassandra L. Smith, *Genomics: The Science and Technology Behind the Human Genome Project*, Human Genome Project Wiley Interscience, 1999.

Charles Staben, *Bioinformatics: A Primer*, Jones & Bartlett Pub. 2002.

Debe DA, Danzer JF, Goddard WA, Poleksic A., "STRUCTFAST: Protein sequence remote homology detection and alignment using novel dynamic programming and profile-profile scoring", *Proteins*, 2006.

Des Higgins (Editor), Willie Taylor (Editor), *Bioinformatics: Sequence, Structure, and Databanks: A Practical Approach*, Oxford University Press. 2000.

Gavin MacBeath and Stuart L. Schreiber, "Printing Proteins as Microarrays for High-Throughput Function Determination", *Science*, 8 September 2000.

Isaac S. Kohane, Alvin Kho, Atul J. Butte. *Microarrays for an Integrative Genomics (Computational Molecular Biology*. MIT Press. 2002.

James M. Bower (Editor), Hamid Bolouri (Editor) *Computational Modeling of Genetic and Biochemical Networks*. MIT Press. 2001.

Luc T. Wille (Editor), *New Directions in Statistical Physics: Econophysics, Bioinformatics and Pattern Recognition*, Springer Verlag, 2002.

Marashi SA, Eslahchi C, Pezeshk H, Sadeghi M., "Impact of RNA structure on the prediction of donor and acceptor splice sites", *BMC Bioinformatics,* Entrez PubMed, 2006.

Minoru Kanehisa. *Post-Genome Informatics.* Oxford University Press. 2000.

Mount D. W., *Bioinformatics: Sequence and Genome Analysis,* Cold Spring Harbor Laboratory Press, 2001.

Nei, M., *Molecular Evolutionary Genetics.* Columbia University Press, Irvington, New York. 1989.

Pavel A. Pevzner. *Computational Molecular Biology.* MIT Press. 2000.

Richard B. Jones, Andrew Gordus, Jordan A. Krall, and Gavin MacBeath, "A quantitative protein interaction network for the ErbB receptors using protein microarrays", *Nature,* 12 January 2006.

Saeys Y, Rouzé P, Van de Peer Y., "In search of the small ones: improved prediction of short exons in vertebrates, plants, fungi and protists", Bioinformatics, 2007.

Schulze Kremer S. (Editor), S. Schulze-Kremer (Editor) *Advances in Molecular Bioinformatics.* IOS Press. 1994

Sensen C., *Essentials of Genomics and Bioinformatics,* John Wiley & Sons. 2002.

Stanley Letovsky (Editor), *Bioinformatics: Databases and Systems,* Kluwer Academic Publishers, 1999.

Steen Knudsen, *A Biologist's Guide to Analysis of DNA Microarray Data,* Wiley-Liss. 2002

Stuart Brown. *Bioinformatics: A Guide to Biocomputing and the Internet,* Eaton Publishing Company, 2000.

Waterman M. S., *Introduction to Computational Biology,* Chapman Hall. 1995.

Index